MESSING
WITH THE
ENEMY

HARPER

NEW YORK · LONDON · TORONTO · SYDNEY

MESSING
WITH THE
ENEMY

SURVIVING IN A

SOCIAL MEDIA WORLD OF

HACKERS, TERRORISTS,

RUSSIANS, AND FAKE NEWS

CLINT WATTS

HARPER

A hardcover edition of this book was published in 2018 by HarperCollins Publishers.

MESSING WITH THE ENEMY. Copyright © 2018, 2019 by Clint Watts. All rights reserved. Printed in the United States of America. No part of this book may be used or reproduced in any manner whatsoever without written permission except in the case of brief quotations embodied in critical articles and reviews. For information, address HarperCollins Publishers, 195 Broadway, New York, NY 10007.

HarperCollins books may be purchased for educational, business, or sales promotional use. For information, please email the Special Markets Department at SPsales@harpercollins.com.

FIRST HARPER PAPERBACKS EDITION PUBLISHED IN 2019.

DESIGNED BY WILLIAM RUOTO

Library of Congress Cataloging-in-Publication Data has been applied for.

ISBN 978-0-06-279599-1 (pbk.)

19 20 21 22 23   LSC   10 9 8 7 6 5 4 3 2 1

For Pepper, I'm eternally thankful you came into my life,
and thankful you will never read this book,
nor have a social media account.
The world is a better place with you,
and was a better place before social media.

Contents

1

Omar and Carfizzi

 Samit al-Muhaajir @SamitalMuhaajir   3h
@selectedwisdom Aslam Awan lived in the UK for 4 years.
telegraph.co.uk/news/uknews/te█.
💬 📄 View summary   ↩ Reply   ⇄ Retweeted   ★ Favorited

"That's him, or at least one of his friends."

I stared at my phone on a wintry Boston night in January 2013, scrolling through my Twitter account. For the past several years, I'd watched terrorists chat with each other and pundits banter and commiserate on my feed, expressing their views on the current state of global jihad. Rarely did the two communities converse, instead expressing themselves in parallel worlds.

"Aslam Awan lived in the UK for 4 years," the tweet read, tagging my handle, @selectedwisdom, and linking to a *Telegraph* news article. The name Aslam Awan meant nothing to me. But someone in Somalia had sought me out on Twitter to bring this story to my attention. I had a sneaking suspicion that it was Alabama-born-and-bred American terrorist Omar Hammami, someone I'd been monitoring for years, or one of his close associates.

Like many other counterterrorism experts, I opined on the travails of al-Qaeda and its followers on a blog. Mine was called Selected Wisdom.com and had an accompanying Twitter account. By 2013,

I'd just finished my second stint at the FBI, counterterrorism had become a hobby rather than my job, and the blog was a distraction from the daily grind in Boston.

Writing on my own meant that I could focus on what I deemed important, pressing issues, rather than the pet projects of poorly informed bureaucrats checking their time in the Pentagon or narrow FBI investigations completely devoid of the bigger picture. On the internet, too, I had an audience. Far more people read what I wrote in late-night blog posts than had ever consumed the analysis I did for the government. Serving as a contractor for countless Defense Department and intelligence community projects, I had written up massive research studies that littered the shared drives of the U.S. counterterrorism community. If they were read, they were routinely dismissed or forgotten amid an endless sea of similar reports pontificating on the strengths and weaknesses of Osama bin Laden and his legions. I don't really blame the U.S. intelligence agencies or the special operations folks scattered around the world for ignoring the countless tomes of analysis they'd purchased. They received far more content than they could ever possibly read, comprehend, and utilize.

Freed from the bureaucratic constraints of Washington, left to my own devices, and with the internet as my playground, I could broaden and deepen my study of terrorists. I had access to many more sources, could choose what I studied, and, most important, could actually engage with the enemy. For lack of a better word, I could *mess* with extremists half a world away—observe their debates, gauge their commitment to terrorist principles, and poke them with queries—all from a laptop at home.

The World Wide Web is and will remain the fastest and best way to be in touch with America's enemies. All of them go to cyberspace to connect with one another and attack us. The internet provided a virtual safe haven for al-Qaeda, its affiliates, and its splinter groups. As terrorists flocked to social media, analysts and researchers outside him into speaking roles at Shabaab rallies and in social media videos. Each public appearance suggested to potential recruits that not only could a Westerner come and join the group, but they ultimately could lead the jihad as well, be the next bin Laden, perhaps. Omar's name was getting bigger, and so was his ego.

The second rule of Western terrorist recruits is that the whiter the recruit, the bigger a pain in the ass he will become for the terrorist group he joins. When terrorists' campaigns sour, divisions grow in the ranks, most often between the idealistic foreign fighters drawn in by online discussions who seek Islamic purity and local recruits motivated more by money, survival, power, kinship, and clan. In 2012, Shabaab's leader, Ahmed Godane, orchestrated a Mafia-style power grab from rival Somalis in the leadership council by publicly declaring allegiance to al-Qaeda without full group consent. He purged the group of challengers through imprisonment or death, all while the group retreated from the Somali capital, Mogadishu, into the country's interior. As is usually the case when times get tough, disgruntled middle managers began to question the boss. Omar, high from his social media following, was principal among these dissenters, having fashioned himself the newest great visionary for global jihad. He offered al-Shabaab's dictatorial emir, Godane, unsolicited constructive criticism and a new vision for mending divides in the terrorist group. Omar, a typical American, thought al-Shabaab's leadership ignored the perspective of its followers and should pursue a more inclusive approach.

Godane received Omar's recommendations like any Somali warlord might from an American in the ranks: as a direct threat to his rule. In a scene that could have been pulled from a *Godfather* movie, Godane followed a playbook he'd used with other challengers. He sent emissaries to retrieve the young American for a follow-up meeting. Knowing he was likely about to be imprisoned or murdered, Omar, a YouTube star of global jihad, went on the run. That was where I came in.

A key lesson from my U.S. Army and FBI time mirrored the famous Wayne Gretzky quote "I skate to where the puck is going to be, not where it has been." I'd watched repeatedly during my first professional decade how ladder climbers ran from one hot topic to the next, seeking fame, fortune, or promotion. I arrived at the Combating Terrorism Center, at West Point, in 2005, and quickly saw how counterterrorism analysts were mostly chasing the latest conflicts in Iraq and Afghanistan. When asked which areas I'd like to research, I thought back to the wise words of one of my favorite Army sergeants: "If you're not the lead dog, you're just sniffing the lead dog's ass." Following the Iraq or the Afghanistan conflict from American shores meant I'd end up making PowerPoint slides for other PowerPoint slide makers who'd deliver them to the lead dog for a cursory glance. *No, thanks—I'll work on something else.*

I was lucky to have a great graduate school professor, Dr. Philip Morgan, at the Middlebury Institute of International Studies at Monterey, who specialized in Africa and had stoked my interest in the region. When the opportunity to study terrorists in the Horn of Africa arose and few jumped at the chance, I volunteered. Mostly by default, I began studying Somalia and its terrorist groups, visited Kenya for some research, and co-authored a large study on the region. In 2010, when I began writing at SelectedWisdom.com, al-Shabaab and its fractious clans featured heavily in my analysis. Over several years, Omar Hammami surfaced time and again in my writing, but one of my Twitter counterterrorism pals, Andrew Lebovich, noticed that Omar had been silent toward the end of 2011.

Only a month after al-Shabaab formally pledged allegiance to al-Qaeda, in February 2012, Omar posted a YouTube video claiming that al-Shabaab had turned on him. Lebovich and I noted this as an unprecedented turn in jihadi circles. Bringing Omar's troubles into the light was an amazing opportunity to undermine al-Qaeda's and al-Shabaab's recruitment of Westerners. We wrote a paper on Omar's troubles, considering possible explanations for how this once triumphant American jihadi hero might have fallen on such hard times at the hands of his own terrorist group.[2]

Throughout the spring and summer of 2012, Omar would pop up randomly on YouTube, pleading for his life. A Pastebin file upload containing his autobiography accompanied one of these videos. The first half of the twenty-eight-year-old's autobiography was the story of his radicalization and recruitment into violent jihad; it was a terrible, unedited volume, but nonetheless revealing. He gave an insider's account of Somalia, Shabaab, and his terrorist acts. Above all, the autobiography provided a first-person account of the disastrous consequences of naively joining a murderous terrorist group in a war-torn country. Hammami's narrative described al-Shabaab as a deeply fractured organization lacking in resources at the foot-soldier level and rife with mistrust. Omar acknowledged that foreign fighters questioning al-Shabaab leaders were murdered by their peers.

Omar's revelations were an information coup for those wanting to dissuade others from joining terrorist groups. I combed through Hammami's dribble and put together my own listicle, entitled "6 Reasons NOT to Join Shabaab: Courtesy Omar Hammami." I meant it partly as a joke, but also as a way to seriously use Omar's own words to explain to potential terrorist recruits why they might reconsider joining. These included Hammami getting a waterborne illness, his poor treatment by trainers and fellow foreign fighters, and the possibility that he was about to be murdered by his own boss. My Hammami listicle got some good laughs from the counterterrorism punditry, and the usual online terrorists gave me some guff, but then I got a phone call.

"You know he reads your stuff." J. M. Berger, one of the world's foremost experts on extremist groups of all types and sizes, had been chatting with Omar Hammami on Twitter via direct messages.[3]

"Really?" I said. J. M. described how Omar, despite being on the run, still found time to tend to his narcissism by Googling himself and evaluating all the positive and negative press he received in the international community. Terrorists, much like modern U.S. presidents, ironically, have a high proclivity for narcissism. But this still struck me as shocking. The writing I did in my free time, pecking away on my laptop, was being read almost instantaneously by an American terrorist trying to escape an al-Qaeda affiliate.

After a few days, I forgot about Omar almost entirely. I was busy; I had a day job, bills to pay. Omar went mostly silent on social media as well, so my writing about him ceased. That is, until I received the Twitter post on the evening of January 5, 2013. The Awan tweet, along with some other social media pings, suggested that he might become active again. Sure enough, @abumamerican returned to life shortly thereafter, ending several quiet months. Omar was alive, and he wanted to talk.

It was late at night, and I was watching my phone. I had one of America's most wanted terrorists itching to tell the world his story. I knew he needed attention. He craved attention. He loved it.

I'd spent the past decade in counterterrorism: two FBI stints, dozens of U.S. special operations projects, research in the Horn of Africa not that far from where I suspected Hammami was hiding out. I'd had hours and hours of counterterrorism training. Seminars in interview and interrogation tactics. Thousands of hours poring over academic studies and terror group communications. I'd submerged myself for years in intelligence analysis drills and taught many sessions of it to students. Now I had a terrorist who wanted to talk to me, and no government bureaucrat telling me I couldn't. I was trained, poor, free of interference, and bored. No one could tell me what to do, or, more important, what not to do.

Okay, Omar, you want to talk, let's talk! I hope you know what you are getting yourself into, because I know what I'm going to do. I've been practicing and waiting for a very long time.

They called it the "gloom period": the interval after the cadets' return from the Christmas holiday and before the reprieve of spring break. Freezing temperatures amplified by howling northern winds funneled through the mountains and down the Hudson River, pummeling West Point's fortress buildings. The uniforms, the walls, the ground, the persistent clouds: they were all gray. If it weren't for cadets slipping and sliding across icy sidewalks as they headed to class, one would not know where the earth met the sky.

We rarely left the compound during the gloom period. It was the toughest time of year academically, and most of us were saving up all the money we had for a coveted spring break in mid-March. We passed endless hours hanging in barracks rooms, dipping tobacco, poking fun at one another, and developing elaborate pranks.

One particularly bitter day, I took a shortcut up a staircase behind the cadet mess hall, my usual route during cold snaps, to minimize my exposure to harsh winds. I never paid much attention as I scurried up to the top floors, but I reveled in the warmth that day as I made my way slowly up the staircase.

That was when I noticed the directory for the first time. It was a glass-encased board mounted on the wall, listing the names and phone numbers of the key personnel operating campus logistics—official names, titles, and phone extensions. I paused, grinned, and pulled out my notebook and an ink pen. I copied down the entire roster for the West Point Meat Plant, an organization led by a man named Carfizzi.

In the early 1990s, West Point operated in its traditional spartan ways, an isolated bastion less than fifty miles from New York City but worlds away from civilization. Cadets memorized the front page of the *New York Times* for news. After the first semester, each dorm room of two to three cadets was allowed a single radio. Beyond radios, the academy allotted each company of 120 or so cadets one television, and the upper classes—seniors, called "Firsties," and juniors, called

"Cows"—were the only ones who could watch during the week. Telephone contact with the outside world came only if you were willing to spend hours waiting in the basement, where telephone booths buried among the barrack's gray walls provided infrequent communication with the outside world. West Point was a prison, voluntary incarceration for those seeking a commission in the U.S. Army. The very best education one didn't have to buy, and, for those who survived the four years, membership in the world's greatest fraternity: the Long Gray Line.

Internally, there was a local telephone network for campus administration, roughly a half dozen phones for every 120 cadets. Internal phone lines were for official business only, or so they said. But for my friends and me, they were a way to have a little fun, to subvert the authoritarian system governing us.

Prank calls started out simple. We'd use them to dupe our friends into silly mundane tasks—needlessly convincing classmates to sweep the hallways for an impromptu commandant inspection that would never come or assembling teams of "Plebes" (freshmen) to gather packages and correspondence that never arrived. The endless rounds of pranks created such mistrust of the phone networks that those on cadet guard duty never really knew what to think or do.

My roommates and I were evil geniuses, and one was a particularly savvy computer and communications master. One day, he dug through the unlocked telephone network box in the hallway and quickly figured out how to wire a phone for our room, one that piggybacked like a party line on another cadet company's admin-room phone line. We located an old telephone from a basement storage closet, and we were in business. We had our own ghost phone line accessing the entire academy from our room. The fun began.

Most days, my buddies and I congregated in our room around the ghost phone for an hour or so of prank calls. I'd take out the West Point internal phone book and scroll through the alphabetical listings trying to figure out how to aggravate each of the academy's services.

I'd dog-ear the page I'd left off at the previous day, progressing from *A* through *Z*, dialing and ringing anyone who would answer.

At first we'd laugh just when I got someone agitated to the point that they would hang up on me. But that got boring over time. The real challenge came when I attempted to create a real-world outcome from the prank call, what I'd later learn to call a "behavior change" in the influence business. For example, I once successfully convinced West Point's ridiculously expansive, elaborate, and effective snowplow operation to attempt to clear a large parade field in anticipation of President George H. W. Bush's helicopter landing that same morning. (No such visit was planned.)

I figured out that the more I made the prank fit the organization I was targeting, the better the effect I could achieve. That's why the meat plant operation's organizational chart I discovered in the stairway was so valuable. The names, ranks, and phone numbers provided me needed reconnaissance to make my pranks seem plausible, generating real outcomes.

"Carfizzi!" I shouted into the phone, my first of what would be hundreds of phone calls to this number over the next several years.

"Yeah, who is this?" a gruff, grumpy older man retorted on the end of the line. I'd connected with the head of West Point's meat plant operation, and in just a few words I knew he fit my needs precisely. His speech, his affect—it was angry, tired, frustrated. Perfect. This was going to be just perfect.

"You know who this is. This is Perez." I impersonated one of his shift leaders from the board. It was a gamble, but if it worked, I knew I'd be far more successful at making this guy crazy.

"Perez, what are you doing?" Carfizzi seemed confused, but he didn't discount me entirely.

"You know what I'm doing. We need more chicken patties up here. We're short, so get them up here fast." I knew that chicken patties and corn chowder, a cadet favorite lunch in winter, were on the menu that day.

"What are you talking about?" Carfizzi said.

"Shut up, old man, I'm tired of covering for you." I challenged him in a way a subordinate should never talk to his boss.

"Hey, you can't talk to me like that," he said.

"I'll talk to you however I want—everyone knows I run the show. Now stop messing around; get those damn patties up here or I'm going to tell everyone the truth about who runs this place." I hung up the phone and we all burst out laughing, assuming that Perez would receive a talking-to later from his frustrated boss.

The next day, same time, I dialed Carfizzi back, impersonating another worker at the meat plant. Same result: he went for it. Several days went by, then I hit Carfizzi at the same number and impersonated another name from the organizational chart I'd copied down. After a week or two, I'd burned through all of Carfizzi's employees and I returned back to the top of the list.

"Carfizzi!" I shouted into the phone.

"Who is this?"

"Perez!" I shouted back.

"Oh, really? Because Perez is standing right here next to me." The game was up for that day.

"Shut up, old man—I'll kick your ass," I trolled Carfizzi a little, knowing his blood ran hot, that any challenge to his seniority and strength drove him crazy.

"You better hope I never find you." Carfizzi had had enough. I heard the frustration in his voice and knew I was under his skin. I hung up. Over the next several weeks, I continued calling Carfizzi, but the calls became routine. He'd answer, get angry, and quickly hang up. I needed a new route to disrupt him. I developed an approach I called the reverse prank call.

Academy staff left like clockwork. By 6 p.m., West Point offices outside of the central cadet area were shuttered. All calls went straight to answering machines, which gave me a new method for driving Carfizzi crazy. I went down the academy phone book and called every number and left a message. On each message, I impersonated Carfizzi and left his phone number, always claiming that I needed to be contacted immediately. The health clinic, janitorial services, animal control, sports ticketing office—Carfizzi didn't know it, but he had reached out to everyone on the base over a couple of nights, and during the daytime, I assumed, each one of these loyal civil servants was returning the desperate plea I left under his name. The entire reverse-prank-call operation took about a week to complete and numbered just shy of a hundred phone calls. Carfizzi had to answer about the "wild dogs running around the meat plant" and the "traumatized cadets" who fell ill from spoiled hamburgers. The snowplow operator may have shown up to clean off the loading dock, and the academy marching band followed up on a request to play at his retirement ceremony. I concocted dozens of bogus stories on these answering machine messages. A week later, I dialed my favorite number.

"Carfizzi," I shouted.

Carfizzi responded, angry and almost saddened. "Why are you doing this to me?"

\\\\\\\\\\\\\\\\\\\\\\\\\\\\\\

Carfizzi had a point. Why was I doing that to him? Prank calls were childish, devious, not right, but wonderfully fun. For me and my buddies, they provided an outlet for challenging the authoritarian system that denied us a life, a way to fight back against our oppressors, a telephonic insurgency against "the man." At this point, I imagine you, the reader, might be a bit confused and disappointed, and have some questions. Yes, I'm sorry I prank-called Mr. Carfizzi; well, kind of sorry. Carfizzi, regretfully, took the brunt of my shenanigans, but he ultimately became the most useful military training I received at West Point—an informal course in social engineering, the backbone of cyberattacks and influence operations. Carfizzi provided me with essential practice to mess with my enemies—the hackers, terrorists, and Russian trolls I came to track on social media. Shenanigans helped me understand and investigate how bad people use ostensibly good mediums to do terrible things.

I reckon you are also wondering how I went from prank-calling academy staff to tweeting with terrorists. How did I go from being an Airborne and Ranger School graduate and infantry officer serving in the 101st Airborne Division to tracking terrorists in a coffee shop in Boston? How did I transition from being a short-lived FBI agent on a joint terrorism task force to briefing the Senate Select Intelligence Committee about Russian interference in the U.S. election? Well, it wasn't by design—it doesn't really make much sense, even—but somehow, I ended up being exactly where I needed to be when I needed to be there.

Growing up in Missouri during the Cold War, we played four things in our neighborhood: baseball, basketball, football, and war. I lacked the height, weight, speed, and talent to make it in the first three, so, beginning in my middle school years, I focused on the last of those. After watching *The Right Stuff*, I became obsessed with military aviation. I researched and memorized the specs of nearly every combat aircraft in history. In the cornfields behind my childhood home, I'd watch McDonnell Douglas test flights of military jets race through the river valleys and dream of becoming a fighter pilot. That is, up until I found out my vision sucked, disqualifying me from any pilot duty.

I went back to the library and started reading every book they had about warfare. I could recite the history of World War II—both European and Pacific theaters—almost from memory. The Vietnam War came next. From there I consumed every book, movie, and television show imaginable on the Cold War, and this led me to tales of spies, espionage, the CIA, and the FBI. I loved them all, and I couldn't wait to leave the suburbs of St. Louis and go serve my country. By high school I'd set my sights on West Point, and I announced to my mom one day in the kitchen, after reading about the Army Rangers, "I'm going to be an Airborne Ranger in the 101st Airborne Division." A decade later, I reported to Fort Campbell, Kentucky, and did just that.

I loved the Army and the infantry, and I did pretty well. But most of my accomplishments in the infantry came not from being a great "grunt," but from being a better intelligence officer. In school and at home, I enjoyed learning about Soviet mechanized doctrine, insurgents in El Salvador, and the mujahideen in Afghanistan. After the September 11 terrorist attacks, I thought I'd soon be leading an infantry company in Afghanistan, but instead I got a call from the FBI. I'd dropped a recruitment card in the mail in August 2001, and after the al-Qaeda attacks, the Bureau started hiring like they'd never hired before. The FBI lowered the bar for entry, and I slipped right over. I left for the Bureau in 2002, became a special agent on a joint terrorism task force in Portland, and . . . I hated it. My first stint in the FBI ended in 2003. I enjoyed counterterrorism, interview, interrogation, and source development, and served with a great squad of investigators who taught me a lot in a short time, but I could not stand the bureaucratic culture and particularly one senior boss. The FBI and I share equal responsibility for my quick departure. I'd transitioned from military life to civilian life too quickly, and the FBI in 2002 and 2003 was a bit of a shambles post-9/11.

Disappointed over my first trip in the FBI, I set off for California, and, after a couple of months of substitute-teaching middle school pre-algebra, I began graduate school in Monterey. The entire time I sat in my international security and development classes at the Middlebury Institute of International Studies, I thought, *Man, I wish I had known these things in the Army and the FBI.* After completing classes in 2005, I landed back at my alma mater and the relatively new Combating Terrorism Center at West Point (known as the CTC; not to be confused with the real CTC, the Counterterrorism Center at the CIA, which hunts terrorists around the world). The hybrid military-academic center bridged two communities: counterterrorism practitioners and civilian strategic thinkers from across the globe. We got to look at declassified al-Qaeda documents, build the best team in counterterrorism, interface with the entire global counterterrorism community, and even teach cadets—a fantastic setup for any academic. In yet another bizarre twist, a senior executive from the FBI stopped by the CTC at West Point looking for some counterterrorism instruction for his agents. A bit shocked to learn I'd been an agent less than three years earlier, he looked beyond my rapid departure and brought the CTC at West Point to train members of the FBI's Counterterrorism Division, and me back to the FBI.

In 2006, I walked back into the exact same FBI classroom in Seattle where I'd taken my first new-agent entry exam less than five years before. With a team of academics and military experts, I started my second stretch in the FBI supporting a range of counterterrorism and intelligence programs at FBI headquarters and the FBI Academy in several periods culminating at the end of 2012—none of which I'm at liberty to discuss nor will I mention in this book, by the way. (Sorry, this isn't a super secret FBI Agent tell all book. That's what we have the Navy SEALS for.)

Thanks to a great leader and some excellent assignments, FBI round two proved far better, much more of what I thought it would be like the first time around. During this period, when not at the Bureau, I supported a range of military and intelligence projects generally referred to as influence operations, part of what is often touted as the "war of ideas." The goal in each was to stem the growing tide of young men joining al-Qaeda and, later, the Islamic State. Each year, their recruitment moved more and more to the internet and social media, and I became engulfed in a series of U.S. and international efforts to counter violent extremism, known affectionately by the acronym CVE.

I'm a high school career counselor's nightmare. I work like crazy to get the most coveted position, and then I quit and go do something entirely different. My supporters think I've had an interesting career, and my enemies think I can't hold down a job. The only two constants during my professional career have been curiosity and contempt for bureaucracy. When I learn new skills or read something intriguing, I don't say, *Oh, that's interesting.* I think, *How I can use this?* I read as much as I can, and I find myself drawn to books about personality types, behavioral economics, and machine learning as much as militant Islam and Soviet espionage. If I'm not learning and the daily work gets routine, I quickly get restless. I'm not disrespectful—I do what I'm told, follow the boss's orders, and hit my marks. But internally, my mind drifts and I start trying to figure out new ways to do something, or dream up elaborate pranks on my colleagues.

The Department of Defense, the FBI, corporate America—they're all great, and I can usually survive there for a bit, but ultimately I've got to get out and go mess around. If you need someone to gin up an insurgency or counterinsurgency, at home or abroad, on short notice, I'm your guy. But if you need me to run a metrics-driven delivery service for a decade, stick to a nine-to-five schedule of conference calls till retirement, I'm out. I need problems to challenge me, keep me calm and content. When I get onto something, like analyzing social media conversations, it's not so much that I want to do it; it's that I can't *not* do it. I'm not sure if it's a personality flaw, a gift, or both. I'm a pain in the ass, but a hardworking one with a mission, and underneath my sarcasm I'm legitimately concerned about terrorism, threats to democracy, and technology tearing up society.

Just to clarify, before conspiracies emerge: the social media experiments I detail here in this book were performed at my home, in a coffee shop, at the gym, on the train, using a laptop or a cell phone. My conversations with terrorists on Twitter, my battle tracking of al-Qaeda versus ISIS on social media, my mapping of Russian trolls—they're research-and-development projects for my mind. The U.S. government didn't tell me to do this; I chose to do it on my own. I've learned in the past twenty-plus years in national security that I'm a lot poorer outside the government, but much happier and in many ways more dangerous. On the outside, I get to pick my own team and network with the world's best, some of whom will make appearances in this book. I'm thankful for having the opportunity to serve my country, and still do, but I now prefer doing it on my own terms, which gives me much greater satisfaction and, I think, has a much more significant impact.

\\\\\\\\\\\\\\\\\\\\\\\\\\\\\\

The internet brought people together, but today social media is tearing everyone apart. By sharing information, experiences, and opinions, social media was supposed to support free societies by connecting users who could collaboratively work through their differences—or so we were told. Initially, we laud each of these new platforms—YouTube, Facebook, Twitter, LinkedIn, Telegram, Instagram, Snapchat. Over time, though, each of these applications ultimately introduces unimagined negative outcomes. Not long after many across the world applauded Facebook for toppling dictators during the Arab Spring revolutions of 2010 and 2011, it proved to be a propaganda platform and operational communications network for the largest terrorist mobilization in world history, bringing tens of thousands of foreign fighters under the Islamic State's banner in Syria and Iraq.

As social media platforms hit their stride, gaining sizable market share and audience engagement, bad actors move in to abuse the system for their own ends. Criminals, terrorists, and nation-states sour human interactions in what were initially wonderful virtual sanctuaries.

Social media becomes anti-social media as the most motivated and best-resourced evildoers learn the strengths and weaknesses of each platform and turn the application to their advantage. Eerily similar to my prank calls at West Point, bad actors use social engineering—psychologically manipulating others to perform actions or divulge confidential information—to force people to unwittingly share their passwords, log-ins, identification numbers, bank records, and personal secrets. Criminals use this personal information to take people's money, terrorists to strike fear in their adversaries, and nation-states, like Russia, to warp the minds of their opponents. Nowadays, hackers and propagandists don't simply seek fame or money, but instead to influence an audience—to create a behavior change among those they target—hoping they might perpetrate a terrorist attack, vote for a candidate sympathetic to a foreign power, or maybe even ignite a civil war. The practical jokes I played twenty years ago are child's play compared with the rapid cycling of schemes attempted on everyday users.

I'm not surprised the #TrumpTrain overtook the American Republican Party, as it looks eerily similar to how the Islamic State overtook al-Qaeda on social media. My Twitter conversations with Omar revealed not just a new generational undercurrent for global jihad, but a wave of social-media-enabled zealots upending not only terrorist organizations but democracies. The quest to pull back the curtain on conspiracy, real or imagined, has pushed social media users to fall back on their biases, believe the impossible, overlook the obvious, and turn on their friends, family, and fellow citizens. None of this compares with what Vladimir Putin's propagandists have been able to achieve in such a short time: a system of mind manipulation that authoritarians now duplicate, and, if left unchecked, it will be adopted by politicians everywhere to overwhelm democratic audiences with waves of conflicting information—fake news—designed to manipulate audiences for a hidden puppet master. The Russians initiated the wave, and now they ride the tide as America's politicians adopt the playbook. Social media manipulation will only get worse as artificial intelligence maps users' thoughts and arms propagandists with unprecedented speed and the power to endlessly amplify their message.

As of now, the future doesn't look too bright. But I have seen successful ways for users—citizens, corporations, and countries—to survive and thrive in our social media. I've seen how some open systems can again be harnessed for good, and how people can protect themselves and their communities. The formula for a social media counterattack against the world's worst is out there; we just need to pursue it, one tweet, post, or share at a time. To find the answer, I propose we start not at the birthplace of social media, Silicon Valley, but in the birthplace of global jihad, Afghanistan.

Three Generations of Jihadi Groups, Leaders & Media

Group	Leaders/ Media Stars	Founded	Location(s)	Media Advancement
Services Bureau (Mujahideen)	Abdullah Yusuf Azzam	1980s	Afghanistan; Pakistan	Print, Audio Tapes
al-Qaeda (AQ)	Osama bin Laden; Ayman al-Zawahiri	1988	Afghanistan; Pakistan; Sudan; Global	Television, Yahoo Groups, Websites
al-Qaeda in Iraq (AQI); Islamic State of Iraq (ISI)	Abu Musab al-Zarqawi	2004; 2006	Iraq	Private Forums, Online Video
al-Qaeda in the Arabian Peninsula (AQAP)	Anwar al-Awlaki	2009	Saudi Arabia; Yemen	Online Magazines
al-Shabaab	Ahmed Abdi Godane; Omar Hammami	2006	Somalia	Social Media
Islamic State of Iraq & al-Sham (ISIS); Islamic State (IS)	abu Bakr al-Baghdadi; abu Mohammad al-Adnani	2013; 2014	Iraq; Syria; Global	Multiplatform Social Media, App Development

2

The Rise and Fall of the
Virtual Caliphate

They were only a few hundred yards from him. Delta Force operators climbed the rocky steeps into the high mountains of Tora Bora, Afghanistan. Just three months after the 9/11 attacks, al-Qaeda's leader, Osama bin Laden, sat surrounded in a cave complex not far from where he'd fought a famous battle in Jaji against the Soviets more than a decade before and later built a terrorist training camp known as al-Masada—"The Lion's Den." Now, barricaded in this high-altitude perch, bin Laden's remaining cohort faced unending air strikes.

Bin Laden sent out a final call to his followers on December 13, 2001: "Our prayers were not answered. Times are dire and bad. We did not get support from the apostate nations who call themselves our Muslim brothers."

Bin Laden's plea proved premature. Delta Force and CIA operatives closed in on his rocky retreat, but they needed reinforcements to seal off the mountaintops. Requested support from U.S. Central Command's (CENTCOM) headquarters didn't come. General Tommy Franks, commander of CENTCOM, instead sought a light U.S. military footprint and chose to use local proxies—Afghan militias—to surround the Tora Bora retreat. This decision provided bin Laden with a lifeline.[1]

Al-Qaeda operatives trapped in the cave complex engineered a cease-fire with the Afghan militia forces surrounding Tora Bora. Some of these Afghan militias even turned their weapons on U.S. operatives as they sought to advance on bin Laden's refuge. The temporary cease-fire opened a narrow window, and bin Laden slipped across the Pakistani border. Al-Qaeda no longer had a physical base in Afghanistan, but it would soon establish a virtual one, which endures to this day.

Bin Laden knew that from Pakistan he could never replicate the al-Qaeda he had founded more than a decade before. Hunted by the entire international community, with his aides and deputies constantly on the run, he couldn't exert the same operational control needed to administer a global terror group. Nor could he inspire, radicalize, and recruit new operatives using the traditional methods he'd come to rely on.

Ironically, in those months following September 11, al-Qaeda seemed to be in decline, but the convergence of two fortuitous events for al-Qaeda reinvigorated the terrorist group. The U.S. calamitously invaded Iraq, fracturing and destabilizing a nation in the heart of the Muslim world. Images of American boots marching into Muslim territory breathed new life into al-Qaeda's ideology. Aspiring jihadis, stoked by al-Qaeda's message, had a focal point for their outrage and actionable goals. They could travel to Iraq to free Muslims from Western oppression.

The internet was the second gift to al-Qaeda operatives now on the run in several continents. Had bin Laden jumped into the Federally Administered Tribal Areas of Pakistan only a few years earlier, only human couriers could have kept the network together. This meager communication method would have quickly failed under intense pressure from counterterrorism forces. While the physical doors into Afghanistan were shutting in 2002, the internet portals connecting al-Qaeda operatives, affiliates, and hopefuls were saving the group.

Al-Qaeda's move to the internet was born not from any grand strategic design but out of the need to survive. The terrorist organization had formed from the remaining foreign-fighters who had answered the call for jihad in Afghanistan against Soviet invaders in the 1980s. Al-Qaeda's strategy from the outset sought to build not a terrorist juggernaut but rather a network of ideologically devout and trained jihadi fighters acting as a revolutionary vanguard for conflicts in Muslim-majority countries far and wide. In the beginning, al-Qaeda's central leadership dispatched operatives as trainers and propagandists all over the world but offered little promotional material beyond training videos and audio broadcasts from ideological leaders. Al-Qaeda's propaganda came not from the center, but from the network. Premier jihadi content of the era came from those on the front lines; videotapes from Bosnia, Chechnya, and Algeria powered the propaganda machine.

Al-Qaeda's first organized forays as a terror group didn't go well. Trainers dispatched to Somalia hoping to mobilize clans to fight against international forces fared poorly. Bin Laden's foot soldiers achieved little, but strategically they learned a good deal. After those missteps, al-Qaeda stepped up its information warfare as a primary method for spreading the word of jihad and inspiring recruits. "Jihadi radio stations operating in Yemen and Somalia will have a more powerful effect on them than nuclear bombs," one al-Qaeda operative argued in a 1994 memo suggesting the regional governments they sought to topple would crumble from within with words more than military might from the outside.[2] Young Somali men, in particular, proved more amenable to stories of jihadi conquest, and al-Qaeda sought to "establish a coordination and communications center to connect the youth in the different areas in and out of the country."[3]

Following this change in strategy, al-Qaeda quickly moved to television to spread its message. In 1997, bin Laden famously gave his first international television interview to CNN's three Peters: Peter Arnett, Peter Bergen, and photographer Peter Jouvenal.[4] But just two years later, al-Qaeda's protectors in Afghanistan, the Taliban, halted bin Laden's major news network interviews. The more attention bin

Laden garnered, the more the world scrutinized the Taliban's harsh rule. To bypass this restriction, bin Laden commissioned a media committee to propagate his tapes to the Arab world. The media committee filmed interviews, crafted exposés, and used a trusted network of couriers to relay al-Qaeda's messages to the outside world—an essential system that helped sustain bin Laden's media presence after the Twin Towers fell.

As early as 1995, bin Laden had also been building an online communications network, using email and bulletin boards. These first centralized efforts to create a digital infrastructure occured alongside those of grassroots groups, students, and religious purists establishing websites promoting jihadism starting in the mid-1990s. Azzam Publications, a London-based publishing house dedicated to promoting the ideas of Abdullah Azzam (often called the father of global jihad), opened in 1996 and posted pro-jihad videos and articles on its website.

Throughout the mid- to late nineties, websites and email chains provided a communications leap forward for terrorists (and the rest of the world), but they had a major limitation: they were one-way modes of information sharing. Bin Laden and his clerics could only broadcast *to* audiences. They could not easily follow up with those inclined to join the ranks. All that changed, though, with the dawn of the new millennium. With the emergence of vBulletin, commercially available software allowing group discussions, and Yahoo Groups, audiences now had a direct window to communicate with Islamist webmasters, clerics, and leaders. In 2001, the Global Islamic Media Front started a Yahoo Group and related website. They required users to acquire a password to access the discussion page. Many others featuring general Islamist discussions with a sprinkling of jihadi talk popped up and down toward the end of the decade. None endured for long before rumors of intelligence operatives penetrating them squelched their dialogue or counterterrorism arrests of forum administrators led to their closure.[5] Two-way communication between al-Qaeda leaders and hopeful jihadis increased, but more content needed to follow to sustain audience engagement.

To fill that void and gain greater control of jihadi discussions, al-Qaeda created an official media group, as-Sahab, which released its first video in 2001, showcasing al-Qaeda's successful attack on the USS *Cole* in Yemen in 2000.[6] Bin Laden recognized the value of ji-hadi websites and began sending audio and written statements from top al-Qaeda leaders directly to al-Qaeda in the Arabian Peninsula's leader, Yusuf al-Uyayri, and his site, Al Neda.[7] Websites and forums served as principal communication points for those around the world inspired by the incredible success of the 9/11 attacks and seeking to join bin Laden's ranks.

The Al-Ansar forum surfaced in 2003 as the first site truly dedicated to jihadi dialogue and activities in Iraq. Al-Ansar lasted only until late 2004 or early 2005, but it signaled two important trends: the rise of password-protected forums and the magnetism of content about al-Qaeda in Iraq's operations. Forums linked disparate supporters in a single channel and built like-minded communities where two-way conversations led to real person-to-person communication and in some cases facilitation into terrorist organizations—bridging the gap between online propaganda radicalization and physical recruitment into terrorist ranks.

Islamists and their more violent jihadi brothers quickly recognized that it would be important to develop as many websites and forums as possible. Having only one website left the network too vulnerable—one site, after all, could easily be permanently shut down. Wise users and administrators sought to replicate and network their online presence. Azzam Publications placed a warning on its site:

> Due to the advances of modern technology, it is easy to spread news, information, articles and other information over the Internet. We strongly urge Muslim Internet professionals to spread and disseminate news and information about jihad through e-mail lists, discussion groups, and their own websites.

If you fail to do this, and our site closes down before you have done this, we may hold you to account before Allah on the Day of Judgment.[8]

Replication of sites and duplication of content became key features of online survival for al-Qaeda supporters. Openly available software and hosting services meant websites and forums could be created by anyone in minutes, and accessed by anyone around the world with an Internet connection. This lowered technical boundary for mainstream internet users meant relatively novice jihadis now had the power to create their own safe havens online.

As websites spread and newcomers clicked into discussions, a second enduring challenge quickly emerged: How does a global terror group maintain security and control over a worldwide movement of unknown people? Password-protected forums were a stopgap measure, and they grew and flourished. A pyramid scheme quickly dictated hierarchy among the propaganda outlets. The forum administrators who had been blessed by al-Qaeda provided exclusive inside information and naturally grew in popularity, garnering more members and more supporters. Competition among forums accelerated propaganda creation and distribution as each jihadi supporter sought to raise his own status in the jihadi community. Forums also created directories where more specific discussion on jihadi subdisciplines could be flushed out and niche content—text, audio, video, and files—could be uploaded and shared.

Al-Qaeda recognized the growing value of forums and shifted its distribution of bin Laden's video productions away from Al Jazeera television to online channels, allowing the content to go viral, spread rapidly, and become more available to jihadi sympathizers.[9] In early 2006, al-Fajr Media Center emerged as the group's official distributor for all things jihad: al-Qaeda, al-Qaeda affiliates, and even non-aligned jihadis.[10]

In those same years that al-Qaeda turned to the internet, so did counterterrorism experts. Political science nerds could now study terrorists from the comfort of home computers, without setting foot in conflict zones. The key to opening these terrorist research troves was simply a password. Arabic-speaking professors fared the best in gaining access to al-Qaeda's forums, but any aspiring counterterrorism expert needed only to team up with a translator and pose as a jihadi supporter to gain a password to an internet forum, and they could be nearly as informed on al-Qaeda's thoughts and dreams as any intelligence agency.

Having quit my short stint as an FBI special agent only a couple of years before, I watched the new era of e-research with skepticism and amazement from the Combating Terrorism Center at West Point. In response to the September 11 attacks, two military academy graduates from rival services—1977 West Point graduate Vincent Viola and 1953 Naval Academy graduate H. Ross Perot—had provided the funding to launch this half-military, half-academic think tank positioned inside the Department of Social Sciences at the U.S. Military Academy. Before the internet, veteran academics spent months and years conducting field research and dangerous in-person interviews to gather insights on terrorists. By 2005, new scholars instead spent days and weeks gathering data online, churning out short analytical pieces nearly as rich in insights.

While bin Laden and company continually improved their e-jihad, two new forces surfaced, weakening their control. The first was Abu Musab al-Zarqawi, a flamboyant jihadi general in Iraq. Online jihadi supporters enjoyed bin Laden's speeches, but they loved Zarqawi and his troops' destructive violence. Zarqawi was a man of action, constantly aggressing against the United States, the UN, the Iraqi government, and even Shia Muslims. Whereas bin Laden and his deputy, Ayman al-Zawahiri, opined from safe havens in South Asia, Zarqawi's Iraqi campaigns executed direct sustained combat against U.S. forces in Iraq.

Al-Qaeda propaganda benefited from a new social media platform, new technology, and a new medium: video. Improvised explosive device (IED) attacks on coalition troops, beheadings of captured Western prisoners like the American Nicholas Berg, suicide bomber operations, and sniper shots on American troops were captured on handheld cameras and rapidly posted online. Links posted on forums could relay jihadi successes large and small throughout the jihadi sympathizer ecosystem. Online, videos of violent actions quickly eclipsed footage of proselytizing al-Qaeda senior leaders. Young jihadi boys wanted action, they wanted blood, they wanted to be Zarqawi. By 2006, foreign fighters inspired by internet postings were flowing into Iraq, seeking to join Zarqawi's legions.

And then there was YouTube, which burst onto the scene in February 2005. Previously, sharing videos required one to upload content to servers and then distribute the links on forums. Servers could be traced and shut down by law enforcement. Digital signatures compromised forum administrators and their online fans, and pursuit by counterterrorists stalled jihadi outlets. YouTube provided a ubiquitous medium any terrorist supporter could access or upload content to. No need to have your own server or hosting service. YouTube was both a spot to upload content and a place to discuss it—the comments sections quickly blew up. Any supporter could access it, and administrators couldn't block their entry by restricting passwords or moderate their speech by suppressing their comments. Furthermore, once content was uploaded to YouTube, the link could be shared on forums far and wide, where other jihadis could rapidly view, share, or download the content.

But bin Laden couldn't quite keep up with social media's rise. Announcements of an impending bin Laden speech from Al Jazeera had become commonplace. Researchers studied the footage looking for clues into bin Laden's thinking and al-Qaeda's direction. Even as a counterterrorism enthusiast, I couldn't get through the summarized translations. His pronouncements went on for thirty or forty-five minutes. He offered little to excite young recruits. He was always perched in the same room and robes, AK-47 by his side. His speech was heavy, weighed down by customary references to past leaders, and he spat out the same narratives over and over. *America is the devil, Muslims must unite and rise up against foreign invaders in (fill in the blank), it's your duty as a Muslim, and one more thing: Death to Israel and the Jews* . . . Who was watching these bin Laden vids? Sure, there's always a devotee here and there engrossed in the words of their prophet. But I couldn't imagine many al-Qaeda supporters spending their time poring over his every word. Aside from the 2005 London transport bombings, bin Laden and his senior leaders hadn't executed a successful spectacular attack in years, and the successes they claimed online came almost exclusively from affiliates, namely their branch in Iraq.

Jihadis *were* watching YouTube, though. It was easy to see they were by monitoring the accelerating number of views for videos showcasing Zarqawi's violent legion in Iraq. The next generation of jihad had already moved on from bin Laden, inspired by a new terror group using a whole new medium. And al-Qaeda in Iraq started its own rebranding in October 2006, changing its name to the Islamic State of Iraq and launching its own media group, al-Furqan. The rebranding and split media operations didn't represent a fracture, at least not yet. Al-Qaeda's al-Fajr continued distributing the newly branded Islamic State's al-Furqan content; it had to. Without jihad's Iraqi legions and their relentless attacks, al-Qaeda's central propaganda outfit had little to cheer.

Forums were still the preferred method of initial content distribution from real terrorist groups and the site of high-level discussions on strategy, but YouTube opened a pathway for commoners to accelerate content sharing and discussion. Content continued bleeding over from forums to mainstream social media applications as Facebook and Twitter overtook the internet.

Forum administrators, partly out of self-preservation, warned in 2009 of the security risks of spreading jihad on social media. But their warnings didn't carry much weight. Jihadi forums weren't exactly safe and dependable, either. Ansar al-Mujahideen shuttered in August 2010 when Spanish authorities arrested a key administrator.[11] Occasionally, many forums would go offline at once, suffering what appeared to be distributed denial-of-service (DDoS) attacks coming from counterterrorism efforts. Every single one of the forums was surely penetrated by intelligence agencies. My colleagues and I joked that there might be more counterterrorists than terrorists on some of the forums.

Even as forums joined jihadi communities together, creating dialogue, content sharing, and engagement among al-Qaeda's core supporters, radicalization and recruitment of new members remained limited and principally in-person. Social media changed that quickly. Video content showing off al-Qaeda in Iraq's deadly rampages went viral, and open dialogue on YouTube, Facebook, and Twitter extended the reach of terrorist groups to Muslim diaspora populations worldwide.[12] Not only could terrorists upload their destructive operations on social media, but recruits, fanboys, and suicide bombers could increase their notoriety by posting video manifestos or martyrdom tapes. On some forums, contributors urged members to join Facebook, seeking to open up discussions beyond jihad's elites. The effects of jihadi migration to mainstream social media quickly became clear even in America.

In 2008, five Americans from Alexandria, Virginia, mysteriously disappeared and then surfaced in Pakistan, where they were detained on terrorism charges. YouTube videos inspired the five friends: it was on the social media platform that they met an al-Qaeda operative named Saifullah, who helped guide their travels to Pakistan.[13] Just a year later, the Somali terror group al-Shabaab's rapid rise, fueled by recruitment of foreign fighters in diaspora communities, pointed to Facebook recruitment occurring in Minneapolis, Minnesota.[14]

Social media allowed rising jihadi groups and their supporters to bypass al-Qaeda's central leadership. Now any terrorist group, large or small, could spread its propaganda, connect with its loyalists, and guide in new recruits. Al-Qaeda affiliates with lower technical capability and limited internet access could create mainstream social media pages.

After 2001, Western counterterrorism continued to improve each year, making terrorist travel to fronts in Iraq and Afghanistan more difficult by the day. Bin Laden and al-Qaeda's central leadership suffered unending drone strikes and losses. Zarqawi's death, in 2006, followed by two years of U.S. special operations raids in Iraq withered the newly branded Islamic State of Iraq dramatically, slowly dimming its star. As some al-Qaeda groups crumbled under relentless strikes, al-Qaeda in the Arabian Peninsula, immersed in war-torn Yemen, ascended in the jihadi hierarchy. AQAP brought attacks on the U.S. homeland through concealed bombs carried across the ocean on commercial aviation or by inspiring homegrown recruits in America to attack at home. The message from al-Qaeda strategists in Yemen shifted: *If you can't join us on the battlefield, then stay where you're at and conduct jihad at home.* U.S. counterterrorism officials' fear of internet jihad may have been overplayed initially, but AQAP's new strategy, the advances of social media, and the rise of a new jihadi star, American-Yemeni cleric Anwar al-Awlaki, confirmed their worst fears.

Awlaki's connections to the 9/11 hijackers and his controversial time as an imam in northern Virginia led to his departure for Yemen in 2004. The Yemeni government arrested him in 2006, in connection with an al-Qaeda kidnapping plot. When he was released in 2007, Awlaki's English YouTube sermons spread like wildfire. One popular sermon offered "44 Ways to Support Jihad," encouraging supporters to perpetrate suicide operations in the West, and a separate message called for jihad against America. He operated a blog and MySpace and Facebook pages, engaging directly with jihadi fanboys around the world providing religious guidance and operational direction.[15] By 2010, Awlaki's social media persona had shifted from religious scholar to freedom fighter. Pictures showed him clad in ammunition belts and sporting a rocket-propelled grenade (RPG) launcher. Another American jihadi online propagandist, Samir Khan, ended his popular jihadi blog *InshallahShaheed*, which he authored from his parents' basement in North Carolina, and moved to Yemen, where he teamed up with Awlaki in 2010.[16] Together they produced and promoted AQAP's new English online magazine *Inspire*, providing directions for supporters worldwide to perpetrate violent plots locally.

Inspire traveled through Twitter, Facebook, and jihadi forums at light speed. Awlaki's virtual inspiration soon connected to attacks far and wide. U.S. Army major Nidal Hasan emailed Awlaki for guidance prior to killing fellow soldiers at his Fort Hood army base. Umar Farouk Abdulmutallab, an online recruit in Nigeria who was attracted to Awlaki's magnetism, traveled to Yemen, where he later donned an underwear bomb before boarding a Christmas Day plane bound for the United States. The bomb failed, but the implications were dire.

Awlaki radicalized supporters via mass social media dissemination and directed them in private messages. Al-Qaeda had finally synchronized all of its online operations on social media. Radicalize, recruit, operate, finance, train, direct—everything could be done from afar on YouTube, Facebook and Twitter. Content and narratives moved around the world at a frantic pace, enticing sympathetic supporters, enraging diaspora communities, accelerating recruitment, and cloaking terrorist operations. The internet had saved al-Qaeda from demise, and social media seemed to be restoring its lethality. At least it seemed that way at first, but the social media era would soon take terrorism in an unexpected direction.

\\\\\\\\\\\\\\\\\\\\\\\\\\\\\\\\

Open-source intelligence became a booming business as terrorists moved to social media. Jihadis liked to show off, feeding their egos through Twitter and Facebook posts retweeted and liked by their diaspora communities and fellow fighters. YouTube videos rapidly radicalized some new jihadi prospects, but they also provided a window into terrorist operations for all those seeking to understand and dissect al-Qaeda's operations, particularly those in Iraq. Much like Hollywood directors, terrorist social media video producers had definitive styles they used and locations from which they like to film. Downloading a jihadi video stream allowed one to understand the size of the terror cell, their accents revealed Arabic dialects denoting their homelands, and stray frames sometimes even showed their hideouts in Iraq. Task organization charts outlining a terrorist group's chain of command could be created from YouTube videos, and their screen names tied to email addresses.

By 2006 and 2007, cadets enrolled in the terrorism and counterterrorism classes I instructed at West Point gathered online videos and created their own real-world intelligence projects from the barracks. The U.S. government, initially, was slow to react to this new intelligence opportunity. During unclassified training sessions, U.S. government personnel often shouted down academics during presentations: "Hey, you can't talk about that. That's classified information." During breaks, academic presenters often conducted a hasty Google search and revealed to these analog counterterrorists that their precious secret intelligence sat on the World Wide Web for everyone to see. The advantages the U.S. intelligence community had once enjoyed were slipping away in the open-source world.

A few years before, in 2004, American journalist James Surowiecki had published a book called *The Wisdom of Crowds*, which described how the internet provided a vehicle for crowds to make smarter decisions than even the smartest person in the crowd, working alone, could make. Collective intelligence mined from the internet through crowdsourcing proved effective in three types of decisions: coordination challenges, where groups work together to determine an optimal solution, such as the best way to get to work or travel overseas; cognition calculations, where people involved in a market compete to provide the right answer, such as guessing the winner of an election;

and cooperation networks, where a central system collects information and the crowd then controls behavior and enforces compliance—think Wikipedia as an example.

The internet lowered the barriers for users to contribute their data and opinions in each of these situations and significantly decreased the costs of collecting such information. The advent of the iPhone propelled this phenomenon even further with social media apps rapidly capturing user experiences. Amazon provided millions of product reviews, Yelp's app created restaurant recommendations for any neighborhood, and Rotten Tomatoes guided people to the better movies available in theaters. As I began to further develop my blog, selectwisdom.com, in 2010, I thought crowdsourcing might be a way to harness counterterrorism analysis in the same way it had helped terrorists march forward online.

I started the blog off slow. I'd write a short post on whatever al-Qaeda affiliate had pulled off an attack that week. I'd profile those groups that I knew better than others. I'd showcase the great research of those who didn't get enough attention for their work, predominantly those outside of Washington, D.C.'s thought bubble. I hoped to gain a sizable audience to gather a statistical sample of opinions, pushing for at least thirty contributors but often getting a bit more. When challenged on my analysis, I'd post a survey on Twitter, collecting answers from a few dozen of the world's emerging counterterrorism superstars and occasionally a real terrorist or two, a few sympathizers participating in terrorist forums, or a terrorist fence-sitter debating travel to a far-off battlefield. They'd provide a bit of perspective, too, adding an alternative view from across the oceans. The following week, I'd publish the survey results on my blog and start a new round of dialogue. My readership was light, but slowly building.

I kicked off 2011 with a new terrorism forecast. Expert predictions litter news outlets every New Year, and since I'd been working on a new theory, I placed my bet on January 2, 2011.

"Osama bin Laden will be killed this year!" I offered, with some supporting analysis.

In the previous weeks, I had employed a prediction technique I'd previously taught in government intelligence analysis courses to estimate when al-Qaeda's leader might be killed. Rumors around Pakistan suggested that bin Laden might be moving about, President Obama was itching to leave Afghanistan, and I checked some actuarial tables to see what percentage of people bin Laden's age (he was fifty-three) die naturally. I also assumed that bin Laden, if he were to be killed by a drone or a raid, would meet his demise in the warmer months, when terrorists roamed more freely and special operations forces picked up their pace. The bet was a longshot, but I didn't believe it to be crazy. Americans had hunted bin Laden for ten years; he couldn't hide forever and still keep control of his network.

The prediction, however, was actually designed as a vehicle for crowdsourcing an important question. What would al-Qaeda and the world of terrorism be if bin Laden were no more? I used my New Year's bin Laden prediction to provoke the audience to answer my "Post–bin Laden" survey. My attempts at crowdsourcing this survey failed miserably. Rather than yielding great wisdom or important insights from experts, the results instead returned a pattern of answers of no consequence. "Nothing will change" and "It doesn't matter" became patent answers from the best thinkers in the field, regardless of the question.

My crowds weren't wise; they seemed a bit lazy and dumb. Even more, on Twitter, they were hypercritical. Reviewers of my analysis couldn't wait to tell me my prediction was wrong or that my questions were stupid. Crowdsourcing the future wasn't working as planned.

I assumed from the outset that the crowd I was building met the preconditions for being wise: diverse, independent, and decentralized. Linked primarily on Twitter, positioned in wildly different geographies, sometimes even in conflict zones, I wrongly believed the crowd I polled had diversity in their opinions and independence in their thinking. While physically the crowd was highly decentralized, social media somehow unified their thinking, overtaking any local knowledge or diverse expertise they had previously accumulated. I went back into the research to figure out why this kept happening.

A political scientist at the University of Pennsylvania, Philip Tetlock, had performed the most expansive collective judgment research to date, in his 2005 book *Expert Political Judgment*. Tetlock surveyed hundreds of experts in political thought over two decades. After thousands of questions, he determined that experts, en masse, were no more successful at predicting future events than a simple coin toss. Tetlock observed two kinds of forecasters. Borrowing from a Greek saying, "The fox knows many things, but the hedgehog knows but one big thing," he classified those good at predictions as "foxes" and poorer performers as "hedgehogs." What differentiated the two groups' success was how they thought, not what they thought. Tetlock's foxes were self-critical, used no template, and acknowledged their misses. Hedgehogs, by contrast, sought to reduce every problem to a single theory, were not comfortable with complexity, were overconfident in their assessments, and placed their faith in one big idea, pushing aside alternative explanations. I saw a lot of hedgehogs in my online surveys and occasional foxes. I needed to find foxes to get insights, and to do it, I decided to employ social engineering, the same techniques I'd used in West Point prank calls. I needed to trick respondents to reveal their true tendencies as either a fox or a hedgehog. Once I found the foxes in the pack, I'd investigate their responses as the outliers in a sea of predictions.

Daniel Kahneman and Amos Tversky, two Israeli American behavioral psychologists, provided the social engineering tricks I needed to separate the foxes from the hedgehogs. Over many years of research, they identified a series of heuristics—mental rules people use to make decisions—and noted the circumstances where biases emerged that led to incorrect judgments. These two gentlemen had determined long ago the predictive missteps I observed in my polls. Status quo bias, a belief that tomorrow will most likely look like today, ruled my responses. Loss aversion, a tendency to avoid anticipated losses rather than pursue equally likely gains, filled the results of counterterrorism policy questions. Herding, the tendency of large groups of people to behave the same way and pursue groupthink, drove my social media recruits to the same set of answers.

Armed with Tetlock's insights and Kahneman and Tversky's heuristics and biases, I changed my approach. The next survey sought not the wisdom of the crowd, but the wisdom of outliers—those gems arising when crowds fail. Instead of asking simple yes-or-no questions, I flooded respondents with as many potential outcomes as I could think of, making it challenging for nonexperts to wade through the responses. I cornered novices and less innovative thinkers by playing to status quo bias. Every question had a "no change" response option, surrounded by responses imitating common thinking stripped from Google searches, newspaper headlines, and cable news pundits. With every question, I offered survey takers a comment box or allowed them to craft an "other" response.

The last week of April 2011, I sent out one large survey designed to ignore the majority opinion and instead look for the outliers. I used some of the same questions from January, like "What will be the chief consequence of bin Laden's death?" but this time I added more strategic questions, such as "Over the next 2 years, the largest portion of Persian Gulf donor contributions to extremism will go to . . . ?" After all the questions, I asked the respondents how confident they were in their answers. Finally, I needed to know why people were outliers, so, while the responses were anonymous, I gathered a bit about their education level, foreign travel, professional background, and preferred information sources. In total, more than three hundred people answered the survey. I coded and studied the results, and then it happened.

I woke to an endless string of text messages on my phone on

May 2, 2011. U.S. Navy SEALs had killed Osama bin Laden in Abbottabad, Pakistan. My prediction had been correct, and my Twitter feed of only a couple hundred followers suddenly became more active than usual. For a brief Google search period, news of bin Laden's death brought a world of visitors to my New Year's prediction. My small blog suddenly had an audience, and I had a new opportunity to gather perspectives from a larger crowd.

I quickly updated the survey from the previous week, rapidly polling people regarding the sudden death of al-Qaeda's leader. I gathered another 150 or so responses to this survey; some were the same people I'd queried the week before bin Laden's death. The results of the two surveys, when compared, showed how a real-world shock suddenly shattered experts' confidence in their assessments.

Analyzing the results from before and after bin Laden's death gave me a select group of foxes. Hedgehogs again picked "No change" in response to bin Laden's death and were highly confident in their assessment. All the responses were anonymous, but when I compared the highly confident "no change" responses, they routinely clumped into two hedgehog buckets. Government folks, military, and intelligence types largely shared the same answers to questions repeating what might commonly be heard in the Washington, D.C., beltway. Academics followed a similar pattern, sticking closely to the preferred classroom theories of the Ivy League. On social media, both crowds whined about how long the survey was and how it was a "waste of time." Matching their attributes with responses, I could see the Pentagon bureaucrats and tired academics in the bunch thumbing their noses at each question. In later survey iterations, I could even see how long respondents spent answering questions. The higher their confidence, the less time they spent answering questions, and, not surprisingly, the more wrong they were.

Meanwhile, some respondents choosing an outlying answer defended their assessment in the comment boxes after they answered. "I know everyone is saying . . . but I just don't agree . . . here's why."

These forecasters knew they challenged the crowd, and they wanted to defend their reasoning, oftentimes providing a kernel of their background, experience, or skills to justify their alternative thinking. The social engineering challenge worked: foxes emerged on nearly every question.

I enlisted West Point classmate and former Army colleague John Brennan as a check on my thinking. We had nearly frozen to death together at the U.S. Army Ranger School and had since stayed in touch. A brilliant cadet and a smart intelligence hand, John was as intrigued as I was by why crowds kept missing on predictive bets. We averaged their responses to questions and averaged their attributes, creating four quadrants. Those always picking outlying responses, and with backgrounds far different from the crowd, we labeled "Red Foxes." They seemed to be contrarians and wildly off most of the time. But those with outlying attributes who only occasionally broke from the norm we dubbed "Brown Foxes," and they seemed to be onto something. A few financial experts floated into this batch, and they proved adroit at predicting where al-Qaeda's money flows would go after bin Laden's demise. Regional experts, meanwhile, more clearly understood how al-Qaeda's affiliates responded to the boss's death.

Most important, I now had several dozen outlying hypotheses indicating where terrorists might go with their hero gone. Social media had brought in terrorist recruits far and wide, but for me, on Twitter, it had helped me narrow the field and pick out the outliers scattered around the world who had unique skills and information enabling them to anticipate where violence would be heading. For each region and terrorist affiliate, I'd tag and track terrorists, journalists, and pundits. I used these signals to understand the jihadi stew boiling and reconstituting as Ayman al-Zawahiri, bin Laden's deputy, took command of al-Qaeda. Al-Qaeda's deputies worldwide began plotting, moving, and attacking, and only an outlier here and there, positioned in a far-off land, browsing a jihadi chatroom or a Facebook group—

not the bureaucrats and pundits sipping coffee in D.C.—could see what was coming.

\\\\\\\\\\\\\\\\\\\\\\\\\\\\\\

Those who believed that bin Laden's death didn't matter for al-Qaeda didn't understand the control he wielded over the group's affiliates. Riffling through the records stashed in his Abbottabad compound, the U.S. intelligence community had unearthed communications from al-Qaeda affiliates around the world. The ties weren't always that strong, but jihadi leaders still sought guidance from bin Laden regarding strategy and finances. Osama's communiqués showed him to be isolated, fearful of drones wiping out his top leaders, and most of all cautious, often restraining the aggression of the younger generation.

Bin Laden encouraged affiliates to hold off on their plans to create an Islamic caliphate—the ultimate goal of jihadi doctrine, the creation of a new Sharia state mirroring that established by the Prophet Muhammad. With bin Laden gone, al-Qaeda franchises no longer had to wait. Al-Qaeda in the Arabian Peninsula moved first as al-Qaeda's senior leadership withered in Pakistan. The Yemeni franchise, powered by the influence of Anwar al-Awlaki's e-broadcasts and a string of near-miss bombings on Western aviation, had already become the central node for al-Qaeda's global network. AQAP created a splinter insurgent group called Ansar al-Sharia that sought to build a prototype Islamic mini-state, an emirate, governed by Sharia law. After initially taking ground and instituting their rule, U.S. drones and the Yemeni army fractured this short-lived Islamic emirate. Still, its efforts signaled what was to come.

Al-Qaeda in the Islamic Maghreb (AQIM) soon followed AQAP's model, teaming with a local Sahelien insurgent force known as Ansar Dine. In early 2012, they pushed the Malian army out of Timbuktu, building, for the moment, the largest geographic Islamic emirate in modern times. AQIM again illustrated extremists' desire to produce something tangible, to actually achieve the vision of jihad by restoring an Islamic state governed by Sharia law. The French quickly swept through the Sahel in January 2013, ending this nascent dream.[17] Even if the Yemeni and French armies hadn't uprooted these emirates, neither AQAP nor AQIM attracted enough foreign fighters to become a robust jihadi playground. That would soon change.

Syria had fractured into full military conflict among rebel factions and government loyalists not long after the Arab Spring touched off in 2010. For terrorists on social media, no fight proved more important. A Muslim country, oppressed by a dictator, full of Koranic historical significance, at the intersection of the foreign fighter infiltration routes that had powered al-Qaeda in Iraq only years before, Syria provided all the ingredients for what would become the largest terrorist migration in world history, one amplified and coordinated at light speed on social media.

Sympathetic Muslim populations from Morocco to the Philippines posted pictures of dead Syrian civilians bombed by President Bashar al-Assad's war machine. The Free Syrian Army grew quickly, and foreign fighters principally inspired to topple the mad dictator Assad rapidly assembled in Turkey and began pouring over the border into Syria. Each new recruit to the battlefield began snapping selfies showcasing their transition from online supporter to on-the-ground commando. Al-Qaeda's Iraqi arm, now branded the Islamic State of Iraq (ISI), being always better on social media than its peers, recognized the opportunity and dispatched a deputy named Abu Muhammad al-Julani across the Syrian border from Iraq to launch a new affiliate: Jabhat al-Nusra. Soon after, a hooded and masked Julani competed for influence among Syrian factions, and jihadi social media production teams kicked into full gear. Videos of jihadi boys in Syria flooded forums, and officially branded clips employing up-to-date video editing spread virally on Twitter and descended into Facebook feeds around the world. Each retweet and favorite raised awareness of the Syrian slaughter, and foreign fighters descended on Syria in greater numbers and at faster speeds than anything seen in any previous jihadi conflict.

The sands were shifting inside jihad. As al-Qaeda launched its Syrian arm, Jabhat al-Nusra, the Islamic State of Iraq surged back into the spotlight in 2012. Terrorist veterans and Sunni Iraqi army survivors, captured and detained by American forces, shared cells and yards in places like Camp Bucca, Iraq. Prison breaks and releases after the American withdrawal returned these hardened fighters to the battlefield. Suicide bombings again picked up around Baghdad.

Abu Bakr al-Baghdadi had initially dispatched his deputy, Julani, to form Jabhat al-Nusra in Syria as an under-the-table al-Qaeda franchise seeking to infiltrate the civil war. Baghdadi's ISI had provided Nusra with men, salaries, supplies, and direction. But the spoils of war, namely oil and foreign fighters, were going to Nusra. Julani sought to report directly to al-Qaeda's chief, Zawahiri, in Pakistan, rather than to his former commander and middleman Baghdadi in Iraq. The two groups increasingly collided in eastern Syria. On April 9, 2013, Baghdadi made a bold move straight out of *Game of Thrones*. He announced the absorption of Jabhat al-Nusra into the newly branded Islamic State of Iraq and al-Sham—"ISIS"—adding the Arabic name al-Sham to its moniker to signify the group's expanded kingdom, which sought to include Syria and the greater Levant region. (Note: the Islamic State of Iraq and the Levant—"ISIL"—is the same as ISIS, merely substituting Levant as the English equivalent of "al-Sham.")

Julani feared he'd lose control of his branch and pledged loyalty to al-Qaeda's global leader, Zawahiri, the next day. Zawahiri, far removed in Pakistan, attempted to restore order from a distance. The three al-Qaeda leaders squabbled behind the scenes over who was in charge, and of what turf. Zawahiri wanted Iraq to be Baghdadi's, and Syria Julani's, with each to report to him directly and independently as an al-Qaeda affiliate. Baghdadi rejected Zawahiri's guidance creating "*fitna*"—divisions in al-Qaeda's ranks—that spilled out into social media. Online rumors and my outlier signals suggested that the schism was real.[18] I watched and marked the sides in the Nusra-versus-ISIS social media debate. Signals showed a major split in al-Qaeda's global community.

In August 2013, I drafted a crude map of what I believed were the two competing al-Qaeda networks stemming from Syria. I posted and distributed it on social media.[19] Again I waited for the responses and looked for biases from the crowd. With only a couple of exceptions, analysts herded to the status quo response—al-Qaeda was fine, and these squabbles were not sufficient to stop their rise. Some outliers provided refinements and pointed out essential questions I needed to answer to prove my theory that al-Qaeda was fracturing. The terrorists who followed me and some of their supporters were less certain than the American analysts. One jihadi sympathizer, an outlier I'd sensed perused al-Qaeda's closed forums and networked with real terrorists, suggested I was onto something. He noted dissension in the ranks but couldn't confirm a complete break.

Through the fall of 2013, social media postings showcasing ISIS boomed and relative chatter about al-Qaeda and its Syrian affiliate Nusra waned. ISIS became the hashtag of choice on Twitter, and the black flag of ISIS a preferred avatar on Facebook. In every European country, there were Muslim boys creating ever more Facebook groups, where they shared content and connected with their friends on the battlefield. Like other counterterrorism sleuths on the internet, I began compiling lists of recruits to Nusra and the newly named ISIS. A rough estimate showed ISIS gaining recruits at a rate of three or four to every one for Nusra.

Countries lauded for their Facebook uprisings during the Arab Spring showed alarming terrorist recruitment levels. Tunisia, where the Arab Spring started, showed one of the highest rates of foreign fighter recruitment to Syria. Libyans, free of Qaddafi's reign, showed up to Syria in droves.

ISIS inspiration brought recruits like ants to food. One lost Muslim boy would make the jump, travel through Turkey, and begin broadcasting from Aleppo or Idlib. Soon after, two of his closest Facebook friends followed his path. Months later, by the end of 2013, entire European neighborhoods of jihadi boys surfaced in Syria, forming battalions sporting black flags and speaking their native tongues. Foreign fighters on Twitter and Facebook from countries like the United Kingdom and France numbered well into the hundreds by the end of 2013. Nontraditional recruiting grounds—countries like Belgium, Denmark, and the Netherlands—suddenly became hot spots for ISIS to draw manpower from. Arab foreign fighters had ignited the mujahideen in Afghanistan in the 1980s and sustained further mobilizations. North Africans had shown up big for al-Qaeda in Iraq during the 2000s, adding to Arab legions. And now, in this third and biggest foreign fighter migration, Europeans, connected on social media, landed in Syria and Iraq in unimaginable numbers.

In January 2014, al-Qaeda's civil war began in earnest. Zawahiri rebuked ISIS for its belligerence, announcing that the affiliate was no longer part of the group. His Syrian affiliate, Jabhat al-Nusra, plotted with other Syrian Islamist groups that launched attacks on ISIS territory in western Syria. Jihadi enthusiasts online didn't know whom to support, and counterterrorism analysts who'd underestimated ISIS sought to minimize the scale and significance of the infighting. Around the world, al-Qaeda affiliates cautiously watched and younger terrorists began cheering more and more for ISIS, even when veteran ideologues sided with al-Qaeda and Nusra. ISIS bombed old al-Qaeda emissaries deployed in Syria and repelled Islamist groups aligned with Nusra, displaying stronger manpower and ferocity than their challengers.

I quickly penned an updated analysis and posted some new, crude charts depicting splits in al-Qaeda and ISIS.[20] I dropped the charts onto social media in February and waited. Feedback rolled in from those overseas and from experts monitoring different countries and regions. The split was real and growing, according to one of my favorite jihadi social media boys, @ShamiWitness. Each month, I updated my charts, plotting new ISIS affiliates surfacing to challenge local al-Qaeda affiliates. Líbya, Afghanistan, Yemen—the list continued to grow. In March 2014, using the wisdom of outliers and their social media updates, I posited that ISIS had already overtaken al-Qaeda as jihad's global leader.[21] More analysts concurred with this estimate than in the summer of 2013, but most still repeated claims that al-Qaeda would reign supreme. Some so vehemently rejected the idea that they challenged my claims in public debates, even as ISIS launched a sneak invasion of Iraq.

In June 2014, a jihadi army of pickup trucks swept across the desert, taking town after town with little resistance. Entire Iraqi army divisions, equipped and trained by U.S. forces only a few years before, fled from ISIS without firing a shot. Many were hunted down, captured, and executed by vicious jihadi foreign fighters. In a matter of days, ISIS stormed and seized Mosul, Iraq's third-largest city, with no more than a thousand fighters. Saddam loyalists and disenfranchised Sunni tribesmen spurned by Iraq's Shia-dominated government were among them. Jihadi social media posts went wild promoting ISIS's advance, drowning out any parallel al-Qaeda discussions.

Twitter feeds and jihadi Facebook groups erupted with videos in early July 2014. Abu Bakr al-Baghdadi, dressed in a black robe and turban emblematic of the Prophet Muhammad, ascended the pulpit at the Great Mosque of al-Nuri, in Mosul. Rarely photographed or filmed, Baghdadi broadcast to the world an announcement not heard in centuries. Less than a year after rebuffing al-Qaeda and only months after being disavowed by Zawahiri, he announced the formation of a caliphate: the Islamic State. Stunned and overjoyed, jihadi sympathizers retweeted, shared, and broadcast the triumph.

ISIS became the Islamic State, and jihadi hopefuls rushed to Syria and Iraq, joining in the vision bin Laden slow-rolled but Baghdadi enacted. As the Islamic State continued to expand and conquer Iraq, its social media capacity grew and was more sophisticated than that of any other terror group that had come before. Its media battalion produced sharp videos replicating the style of video games. Rather than bin Laden's dull monologues, new Islamic State spots showcased global foreign fighters engaged in firefights on the front lines, killing their adversaries and instituting Sharia law. Islamic State media savants dubbed and translated messaging in dozens of languages, opening their world to new recruits: disenfranchised Muslims on five continents. While combat videos grabbed followers' attention, footage showing Islamic State governance and daily life provided worldwide jihadi supporters with physical evidence of caliphate achievements never before seen in the modern age. Over the next three years, the Islamic State's al-Hayat Media Center created and distributed more content than al-Qaeda ever had, and its show was far superior.

ISIS member and supporter social media accounts on Facebook and Twitter were everywhere numbering in the tens of thousands. They employed hashtags to signal major releases and even paid spammers to send out tweets on their behalf. They launched campaigns on Twitter in concert with distribution on Facebook and on a host of other social media applications. ISIS also upgraded its social media teams by pairing them with hackers, who formed a kind of technical brigade. The technical brigade worked on hacking and information security for the group. Together, the teams found workarounds to avoid Twitter's shutdowns and maximized support from its online fan base. Social bots promoting ISIS appeared, and ISIS and its supporters worked continuously to avoid Twitter's controls and account closures.

Twitter's closures ultimately became an exhausting battle for ISIS, so the group moved its operations to the social media platform Telegram, whose encryption and more closed network blended the terrorist forums of the old internet with new social media applications. Forced to move beyond standard social media apps, ISIS even tried to develop its own social media platform to communicate with its members and supporters. By 2016, ISIS's digital connections with its global supporters were powering waves of terrorist attacks in Western countries.[22]

When two brothers attacked the offices of *Charlie Hebdo* magazine, in Paris, in January 2015, killing twelve people, they were tied not to ISIS but to al-Qaeda in the Arabian Peninsula.[23] But AQAP appeared unaware of the attack and took a week to produce a rudimentary video using mainstream news footage to claim credit for the attack. Meanwhile, a second attack later that day by an associate of the two brothers pledged loyalty to ISIS. The Islamic State's social media army rapidly took credit for and promoted the *Hebdo* attacks as their own. To common observers on social media, the *Charlie Hebdo* attack likely appeared the work of ISIS rather than al-Qaeda, though it's likely that neither group knew of the plot beforehand.

For al-Qaeda, the internet was its savior, and social media its undoing. Websites and forums sustained bin Laden's control and influence, propagated his message to key nodes worldwide, and suppressed critics challenging his global view. Social media initially seemed to be the next frontier for al-Qaeda as its Iraq affiliate took to YouTube and rallied a new generation into its ranks. Emerging social media applications opened the doors to the masses, spreading jihad far and wide, and out of al-Qaeda's reach. Instead of expanding al-Qaeda's brand, Facebook and Twitter powered competitors, brought in detractors, and unleashed dissenters from jihad's younger generation.

Through their base in Peshawar, bin Laden and company established physical connections, cementing trust and loyalty among adherents. The internet provided a medium for reaching thousands of potential recruits but opened the door to law enforcement and intelligence. Trust, faith, and loyalty remained unknown and untested in this new virtual world. Ayman al-Zawahiri, bin Laden's deputy and now leader of al-Qaeda, privately worried about social media newcomers in communiqués with the boss toward the end of the decade.

Jihad's younger generation held few direct links to bin Laden's veterans, and their preferences didn't match al-Qaeda's wishes. Bin Laden and his secretaries had founded their jihad as an exclusive, ideological elite created through attendance at training camps and hardened on the battlefield. On social media, though, everyone could become their own imam, picking and choosing from sacred texts justifying any act of violence. Social media fanboys demonstrated more commitment to violence than allegiance to al-Qaeda's version of Islam. As the Islamic State overtook al-Qaeda, it diluted clerical authority, opening jihad to the masses via any means necessary as long as it achieved one goal: an Islamic state.

Ayman al-Zawahiri, presumably trapped in Pakistan, has watched as what he championed slipped out of his grasp. The Islamic State, in less than two years, overshadowed its forefathers, achieving part of jihad's vision and leading a violent terrorist rampage around the world far greater in scale than anything al-Qaeda had ever achieved. As expected, the Islamic State's incendiary violence and rapid rise brought an international response, and al-Qaeda affiliates raced to catch up, remodeling themselves in some of the Islamic State's ways. By 2016 and into 2017, Twitter, Google, and Facebook had all taken steps, slowly but successfully, to rid their systems of jihad's most prolific propagandists and operators. Surviving foreign fighters and their remaining fanboys and fangirls scurried to alternative social media applications, seeking encrypted, secret communications channels, where they could regenerate and hope to proliferate. To date, this hasn't happened, and young jihadis openly whine in closed forums about the need for a new social media home. Websites to forums, forums to social media, mainstream apps to encrypted apps—terrorists have and will continue to evolve, migrate, and innovate on these technologies and whatever comes next.

3

"That Is Not an Option Unless It's in a Body Bag"

"A return of the crusades, but the cross could not save him from the sword." In January 2013, I awoke to a tweet from al-Shabaab.[1] A dead French soldier surrounded by his military equipment lay on a bright orange blanket. A gold cross necklace, the kind commonly worn by Christians around the world, had been pulled from his shirt collar. French officials acknowledged the loss of one of their soldiers during a failed rescue mission to free a French hostage held by al-Shabaab in Somalia.

Soon after the pictures surfaced, Twitter closed al-Shabaab's account, @HSMPress, for violating the company's terms of service when they posted violent images. But al-Shabaab would come back to Twitter soon after the closure. While many laud ISIS's use of social media, it was al-Shabaab that mastered Twitter first.

Fresh out of Afghanistan and unwelcome back home in Saudi Arabia, bin Laden's first troops had settled outside Khartoum, Sudan. Itching for a new campaign in a Muslim land, al-Qaeda dispatched the jihadi equivalent of Special Forces teams to Somalia in 1992, hoping to train, recruit, and inspire Muslim foot soldiers to attack United Nations forces who were working toward peace in the famine-stricken country.

But al-Qaeda's first missions didn't fare any better than other foreign interventions in Somalia. Terror group trainers venturing into the hinterlands immediately encountered resource constraints. Food, water, weapons, equipment—the challenges faced by any Somali warlord were also a problem for al-Qaeda. The country's clans were open to the training but were less interested in perpetrating the kinds of attacks al-Qaeda's central leadership wanted. They were more focused on rival local clans than interested in attacking United Nations troops. Over a couple of years, al-Qaeda's operatives achieved little in Somalia's fractious environment, expending resources with little return. "Leave it, it is rotten 'tribalism,'" said al-Qaeda's strategic planners.[2] Similarly, the United Nations' intervention to stem Somalia's chaotic violence abruptly ended after U.S. forces lost eighteen soldiers, with many more wounded, in a protracted battle in Mogadishu, later portrayed in *Black Hawk Down*.

Al-Qaeda left Somalia behind, but its departure didn't stop a jihadi strain from festering. Amid Somalia's endless clan warfare, Islamist groups used violence to secure their turf emerged. Somalis trapped in Islamist-dominated areas tolerated their rigid ideology and severe punishments, in exchange for security and any economic opportunity. The Islamic Courts Union (ICU) formed in the absence of an official Somali state in southern and central Somalia. This alliance of Sharia courts allowed for intertribal resolution of disputes and some semblance of governance in a country stricken by more than a decade of war.

The ICU's reign didn't last long. The Ethiopian army, assisted by the American military, acted swiftly and quickly crumbled the ICU, which fragmented into a more virulent form.

Al-Shabaab was the youth wing of the ICU—its name translates as "the youth." Like most younger generations in militant organizations, its members pushed for a degree of violence that far surpassed that of their forefathers. Shabaab emerged from the ICU's ashes and undertook a vicious campaign against the Ethiopians and Somalia's

Transitional Federal Government. Shabaab's run over the past decade is unique not for its terrorism, its cause, or its actions, but because it used social media to propel its brand better than other al-Qaeda franchises, despite hailing from a country with few computers connected to the internet.

When people imagine Somalia, they recall images of starving children, burning buildings, and warlord gun jeeps—absolute deprivation, medieval times in the modern day. Somali refugees fled the country's civil war in the early 1990s. These asylum seekers scattered among new communities in Europe and North America, hoping for a better life. The Somali diaspora grew in the 1990s, just as the internet was rapidly expanding. Somalis abroad wanted news from their clans back home, and soon diaspora websites sprouted, connecting clansmen from Mogadishu to Minneapolis.

With their home country in tatters, Somali refugees became the economic lifeline for those left behind. A million Somalis outside the country were sending an estimated $1.5 billion per year into Somalia by 2014. Making these fund transfers required telecommunications. Technology adoption in Africa rarely follows the pattern of the rest of the world. The continent's development at times is so behind, it'll skip some advancements entirely—say, landline telephones—and adopt an invention two or three generations ahead. Somalia, ravaged by war and completely ungoverned, has led the continent in cell phone and money transfer companies. Today, Somalis gripped in violence or starving from famine might nevertheless be able to talk to a fellow clansman half a world away and receive mobile money in an instant—a strange twist.[3]

For many Africans, their first connection to global telecommunications was not mediated simply through a computer connected to the internet, but through a mobile phone social media platform. When searching for information, an African might say, "I'll look it up on Facebook," rather than the common Western internet shorthand, "Google it." Somalis started off not on a computer terminal or a laptop, experimenting with a search engine, but with a cell phone, collaborating with their kin and tribe on Twitter and Facebook.

Al-Shabaab's online path in terrorism mirrors the strange technological trajectory of Somalia. Despite their long-standing ties to al-Qaeda, during the early 2000s few terrorist operatives scattered among the ICU and later al-Shabaab could plug into al-Qaeda's internet forums. With internet penetration hovering at less than 2 percent, only elites tethered to urban infrastructure could even access the World Wide Web.

In 2007, as I prepared for a research trip studying al-Qaeda's operations in Kenya and Somalia, a colleague asked me if I'd check into terrorist use of the internet in the Horn of Africa. "I think these guys [Africans] are all over the forums," he said. This assessment was completely off base. I found only a couple of internet hookups when I arrived in Lamu, Kenya, the last major coastal town before the Somalia border and a short boat ride from where al-Qaeda operatives openly resided. Internet connections were painfully slow, and when I attempted to check email, I could hardly log on, as the connection's entire bandwidth was being used to make bootleg CDs of American music. Prospective terrorists in Africa at the time largely weren't networking on the internet.

Shabaab followed the terrorist internet playbook, but the results weren't the same, because it was late to the game. Shabaab's media wing, al-Kataib Media Foundation, produced videos and raised its profile further with support from al-Qaeda's al-Sahab Media Foundation. But Shabaab's websites and videos came online just as global counterterrorists began shutting down terrorist platforms. Luckily for Shabaab, social media lowered the technical challenges to accessing global communities, and it took advantage of the opportunity.

Al-Qaeda in Iraq took YouTube to another level through videos, and now al-Shabaab brought terrorist dialogue to peak effectiveness with Twitter. @HSMPress became Shabaab's Twitter handle on December 7, 2011, and within days, thousands were following the account. In 140 characters, Shabaab reached its followers, the Somali diaspora, global terrorist sympathizers, and even its enemies.

@HSMPress's tweets pushed sharp barbs with a British dialect. Shabaab's Twitter operator was haunting, challenging, and engaging, providing the group's worldwide supporters with battlefield updates noting its successes. These tweets were different from al-Qaeda's pontifications, though. Tweets were in English, not Arabic, the pithy jabs of punks, not prophets. Posts spat out ideology in talking points, not scripture. They engaged everyone: fanboys, friends, and foes.

"Update: Last night's attack on #Taabto lasted an hour & resulted in the deaths of 3 #KDF soldiers. An ammunition store was also set ablaze." The tweet linked to photos of dead Burundian peacekeepers believed to be killed during the fighting. When not detailing battlefield successes, Shabaab's Twitter handle reached out to potential recruits and donors through narratives of Somali nationalism or Islamic duty.

The Kenyans fighting Shabaab from the south took to Twitter in response. An account under the control of Major Emmanuel Chirchir fired back at al-Shabaab, creating the first international terrorist-versus-counterterrorist Twitter battle. The Kenyan army knew that Shabaab moved weapons and supplies via donkey caravans. Major Chirchir's tweets about restricting donkey sales to Shabaab noted that "any large concentration and movement of loaded donkeys will be considered as Al-Shabaab activity."[4]

Shortly after, Shabaab responded with wit. "Your eccentric battle strategy has got animal right groups quite concerned, Major."[5]

Shabaab's tweets had an appeal unlike standard terrorist missives. It felt real, as if someone was listening to what was being said rather than strictly blasting propaganda. Shabaab's account directly communicated to supporters who could join in the conversation in real time, adding to the community appeal and allure of Shabaab.

In January 2013, the @HSMPress Twitter handle vanished after posting photos of the dead French soldier but quickly reemerged under a similar name. The new account's followers were down, but Shabaab was not out of the Twitter fight. Later, in September of that same year, Shabaab live-tweeted its massacre of innocent, unarmed civilians at the Westgate Shopping Mall in Nairobi.

"The Mujahideen entered #Westgate mall today at around noon and are still inside the mall, fighting the #Kenyan Kuffar inside their own turf." @HSM_Press1 broadcast the attack in real time. Twitter was not simply a means of broadcasting the attack but was now an operational platform for coordinating violence. Shabaab's shock troops in Kenya relayed updates and pictures from inside the shopping center directly to Shabaab's leadership. Meanwhile, Shabaab's pronouncements provided a competing message to the news broadcast by the mainstream media.

Westgate foreshadowed a coming trend: terrorists using social media applications not just for propaganda but for operational coordination. In the weeks after the heinous attack, Twitter took a beating in the press. Officials and pundits wanted to know how the social media application could allow terrorists to proliferate, radicalize, recruit, and attack innocent civilians. Twitter remained steadfast in protecting direct communications, but it soon closed down some accounts. The Shabaab-connected Twitter accounts I monitored began disappearing. But it was too late; the damage had been done. Al-Shabaab's exploitation of Twitter for all terrorist activity provided a social media playbook for terrorists worldwide.

When times are good and terrorist groups pontificate on social media, recruits join in droves. But when victories are elusive, terrorist groups find themselves under attack online from those who once supported them. In 2012, social media began turning on al-Shabaab and al-Qaeda. When Ayman al-Zawahiri took command after bin Laden's death, he knew he needed a quick win. Al-Shabaab offered its allegiance in February 2012, and al-Zawahiri accepted. Previously, bin Laden and his former secretary based in Somalia, Harun

Fazul, had avoided formal ties with al-Shabaab. Still chastened by the failures of al-Qaeda's 1992 misadventure—which took place before Zawahiri joined the organization—bin Laden understood that jihad in Africa brought unanticipated pitfalls. Fazul, perched in the Horn of Africa for more than a decade, thought that al-Shabaab's young, violent boys lacked the ideological understanding to officially be part of the terror group. For al-Qaeda, Somalia provided a better safe haven than al-Shabaab made a good partner. Soon Zawahiri learned why bin Laden and Fazul had hesitated in accepting Shabaab publicly into the fold.

Ahmed Godane, leader of al-Shabaab, pledged allegiance to al-Qaeda in 2012 and in so doing began the long unraveling of the Somali terror group. Godane issued the pledge without consensus from all of Shabaab's leaders. A schism in al-Shabaab soon appeared. Clans broke from the group. Shabaab fractured and retreated into interior Somalia, and a threatened Godane began killing off rivals. Throughout this period, al-Qaeda's foot soldiers hiding among Shabaab fell one by one to airstrikes. Fazul, a top wanted global terrorist and confidant of bin Laden's who had a legendary track record for evading counterterrorists, suddenly died at a Mogadishu checkpoint, presumably one he'd driven through many times.

In Somalia, rumors swirled that Godane had tipped off counterterrorists to help them kill old al-Qaeda hands—terrorists were killing terrorists. And on Twitter, Godane's power grab brought backlash he never expected. Shabaab, once a master of Twitter, would now become its victim. Those fleeing Shabaab's ranks were embraced by Shabaab's enemies, and on Twitter, unlike terrorist forums, nothing short of death could moderate their speech. Remember Omar Hammami, the American-born terrorist? He talked back to Godane during this schism. When he went on the run from Shabaab's assassins, I connected with him on Twitter, and our conversations became a social media weapon against the terrorist group he'd once promoted.

〰〰〰〰〰〰〰〰〰〰

Terrorists want to talk for a reason. Most of what Americans see or hear regarding al-Qaeda is just propaganda—some truth mixed with exaggerations and falsehoods. Bin Laden's long speeches, Zarqawi's impassioned machine gun videos from the front lines in Iraq—they're all staged, prepared promotions designed to motivate lost Muslim boys to travel to a fantasy jihad where they'd give their life for no real purpose.

On social media, though, those in the West studying al-Qaeda were able to peek into the lives of real-life terrorists. The vast majority of terrorists on Twitter hoped to tell their story and promote themselves under bin Laden's banner. Jihadi boys and their leaders principally served their egos as much as their faith. If there weren't martyrdom videos, there wouldn't be too many suicide bombers: no reason to die in a far-off land if no one knows about it.

Hammami and his small band of minions talked for a different reason. They were in trouble.

〰〰〰〰〰〰〰〰〰〰

Rapport building comes first. That's what you learn at FBI interviewing and interrogation training at Quantico. Talk to them as a peer; place your body in an open position, mirroring the subject's body movements. Show interest in their life and well-being; help them relax so as to prepare for the real questions that you truly want answered. Only after building rapport do you go for the goods.

Good interviewers, whether they be special agents or journalists, know what information they want before they start the conversation. When the subject is comfortable, their guard a little down or a bit more trusting than they should be, that's when the real questions come out or, better yet, the subject reveals key points without prompting. I didn't want to engage Hammami; I wanted Hammami to talk to me. I didn't have the luxury of seeing him and chatting, mirroring him physically to ease his apprehension. Instead, I had to rely on 140 characters to grab his attention, build rapport, and nudge him along.

Before engaging Omar on Twitter that night in Boston, I slipped a piece of paper out of my office printer. I outlined my goals for publicly chatting with him, sketching out three objectives. I'd watched other counterterror analysts and pundits try to chat with terrorists online, and they seemed to strengthen rather than weaken their opponents. Most sought to tell terrorists that their "ideology is bad" or engaged in theological debates regarding Islam. For terrorists who've traveled abroad and devoted their lives to pursuing an extremist ideology, these discussions provided public sparring with infidels, where the terrorists could showcase their Islamic mastery, affirm their faith, and grow their brand. But Omar was a publicly disavowed American foreign fighter on the run from his own terrorist group; I sought something different from him.

First, Omar needed to keep talking: the more he talked, the more he illuminated himself, the terrorists around him, and future foreign fighters contemplating travel. Each tweet provided information that would reveal his true location, friends and foes, strengths and weaknesses. Armed with that data, I'd write summaries of his activities. My posts on his adventures had caused him to engage me; I hadn't engaged him. I wanted to continue this trend; I knew his desire for attention would be his undoing.

Next, I wanted more information about the infighting among terrorists. Omar enjoyed describing how members of al-Shabaab failed to follow the strict Islamic law they professed. I wanted him to continue crapping on his old overlords. I wanted him to appear as American as possible, for him to seem like a foreign troublemaker to other members of al-Shabaab. The more Omar talked with Americans like an American, the more he discredited himself in jihadi circles and undermined his appeal to Western foreign fighters' desiring to join Shabaab and al-Qaeda.

My last objective, and definitely the toughest one to achieve, was to get Omar to admit that al-Shabaab and al-Qaeda were on the decline. No aspiring jihadi hopes to be the last recruit in a failing terrorist group. Omar was a terrorist, he had been in the fight, and a Western turncoat's words carried far more weight with potential recruits than any American counter-radicalization program. If he trashtalked al-Qaeda, terrorists in the ranks and future terrorists waiting on the sidelines would listen.

My first action, before pursuing any of these objectives: verify Omar Hammami as the real person behind the @abumamerican Twitter account. Omar had been on the run for many months, posting intermittently from the account, but journalists and naysayers who couldn't get Hammami's attention claimed that the account represented the work of an impostor, a young jihadi wannabe stereotypically sitting in a basement punching on a computer. The signaling tweet on January 3, 2013, proved to be the harbinger I was expecting. Omar came to life on January 5, and this time I was free to mess with him.

Hammami's Twitter kicked off 2013 like an insider's tell-all book broadcast two sentences at a time. Each Omar tweet brought another salacious complaint against al-Shabaab and its dictator, Godane. Hammami claimed that Shabaab's leaders reaped the spoils of war, leaving nothing for the people, and allowed the sale of the regional drug qat so as to tax a substance otherwise illegal under Islamic law. He went further, explaining how old al-Qaeda hands, like bin Laden's former personal secretary, Fazul, had been killed off by Shabaab's emir.

"Fazul was told if u meet up with mukh robow u'll become a bbc headline. He did so then he did become." [Referring to Sheikh Muktar Robow, a deputy leader of Shabaab and rival of Godane]

"Being aq was shunned until the last real op, fazul, was out of the pic. Then being aq became an oblig to be killed for." Hammami asserted that Godane had avoided al-Qaeda until local challengers were killed off. Then, with Shabaab securely under his rule, Godane pushed for a formal, public alliance with al-Qaeda strengthening his personal connection to the terror group's senior leaders in Pakistan and solidifying his control over Shabaab in Somalia.

Hammami also explained why his plight over the past year needed to be posted on YouTube and discussed on Twitter. Jihadi forums—the debate-and-distribution platforms where real jihadi terrorists congregated online—censored him.

"Amriki gave advice as amiir of the haraka's lajna shura, they refused. He tried scholars, aq yemen, aq central, but nothing."

"He tried the jihadi media mafia but found them living in never-neverland."

Omar had been blocked from those very discussion boards that had helped inspire his radicalization and recruitment in Cairo before landing in Mogadishu. But on social media, no jihadi administrator could censor his rants. Curiously, Omar's Twitter handle spoke at times in the third person, referring to Omar as "Amriki," an Arabic transliteration of "American." Arabic regional monikers were a common jihadi way of describing foreign fighters and their homelands, but it was odd for someone to refer to himself in the third person. Unless, of course, Omar wasn't alone or was relaying answers to someone who managed his Twitter account from another device.

That night, I powered up my laptop and began writing a blog post at Selectedwisdom.com detailing Hammami's new disclosures. I would amplify and spread awareness of Hammami's plight to both terrorists and counterterrorists, but, more important, I'd offer something for Omar to click on. I needed proof of life, that it was truly the American terrorist. I posted my Omar summary and went to bed, then woke up quite pleased with the outcome.

"Clint watts: 'i have a dream' was a black man's fantasy before it became reality. America was a mess b4 it became a menace."

"Clint re post 2: amriki suspects aq central are in his same sit on a grand strat level. To speak openly would be 2 shoot ones foot."

He had read my post. Omar was hooked, and now I knew I had

him visiting my blog. I logged in to my Google Analytics account monitoring traffic to my website, and sure enough, the map displayed several hits from Somalia, all coming from mobile devices. I was one step closer to narrowing down Omar's technical signature. Beyond that, I had a terrorist telling me about the inside baseball of al-Qaeda. He believed that al-Qaeda Central—its senior leaders—weren't happy with Shabaab's behavior. But al-Qaeda's leader, Ayman al-Zawahiri, couldn't make a public statement disparaging its newest disobedient affiliate without embarrassing himself, having ushered in a franchise in Somalia that bin Laden never fully trusted or publicly embraced.

Later that day, Omar dropped a new release. Another video detailing his plight. This one was a rerun, posted on YouTube. He again asserted that al-Shabaab was hunting him. He followed up shortly after with a document posted on the anonymous Pastebin website. The doc provided further details about the treachery inside Shabaab's ranks. Omar narrated how Godane had killed off his deputies, al-Qaeda operatives, and dissenters Mafia style, jailed foreign fighters unjustly, and preyed on locals, particularly from lesser Somali clans.

The document provided a dozen different ways for me to engage Hammami and repurpose his story. For me, it was another late night, but a fruitful one. I read the document and built my strategy. I'd take Hammami's side in every claim he made, mirror his arguments with my own, and do a better write-up. Sometimes I'd ask questions in the blog post, hoping he would read and answer them on Twitter. Other times I'd deliberately get things wrong in his story, so that he would correct me, confirming that he was reading my writing. In both cases, I'd nudge him to publicly engage me so his terrorist supporters would see him palling around with a former U.S. government agent. I closed that night's post like others, noting that, despite all the great information we were receiving from this Twitter account, we still had no idea whether it truly was Omar Hammami and not some impostor. Again, I posted late at night, and the results were almost immediate.

"Clint watts, no one is providing u with info. The ummah and history have a right to know. Then we'll mend our house and rise."

Again, just as in previous posts, the @abumamerican Twitter account called me out by name—exactly what I wanted. Even better, the account had posted two pictures. The first provided the verification to stop my naysayers: Omar, front and center, joined by two other foreign fighters, one a Somali British man and the other an Arab man. In his hands Omar held a piece of cardboard. Scribbled in black was a date, JANUARY 11 2013. *Thank you, Omar!*

The second photo showed Omar atop a large wagon being pulled by his donkey. Married, with kids, Omar fancied himself now as both a purist resistance fighter for Islam and a man of the people. A sign of Hammami's colliding worlds and confusing personas. At times, he'd yearn for his place among the great jihadi ideologues, and during other moments he seemed to yearn for America and Americans. My first step complete, Omar's identity confirmed, I refined my strategy further for future engagements, based on what I saw in these two pictures.

I began looking for those times when the Hammami account appeared to be tweeting in the first person. I didn't mind getting scoops from his pals, but I preferred hanging out with Omar on Twitter, where I'd get more rapid and pure responses. Over the next few weeks, I decided the best time to get good engagement from him was during my early mornings. Half a world away, Omar would be on his phone during the late afternoon or early evening Mogadishu time. Instead of posting very late at night, as I had done previously, I changed my pattern to match his, mirroring his high-engagement times so I could converse with him more quickly. Over the next few weeks, I'd write my blog posts the night before, hold them in draft format, and then head to bed. The next morning, as I fed my daughter a bowl of oatmeal, I'd plop my laptop on the counter and watch the Twitter feed in one panel and my Google Analytics account in a separate window.

Ping! Omar posted his first tweet of the morning. I'd quickly post an article on my blog discussing Omar's plight, then tweet the link out into the Twitter tornado. A couple of minutes would go by, I'd clean the kitchen, rinse some dishes, pour another cup of coffee, and then watch the Google Analytics map showing accounts around the world landing on my web address. Somalia would turn orange, and dots would begin emerging from the Horn of Africa. *Hi, Omar, thanks for visiting.*

Two or three minutes later, Omar would start applauding or correcting my analysis with snarky tweets. I'd then examine the Somalia visitor data on my blog, and in the few minutes since I'd posted, there had been only one hit from Somalia. I repeated this process for a few months, and after several iterations I felt confident that I knew what type of phone Omar used and why he would write my name out in his tweets. Omar could access the internet, but he didn't have apps, nor could he type the @ symbol, which his device didn't have. Without being able to write my Twitter handle, he had to get my attention by spelling out my name.

I didn't hope to capture Omar Hammami, nor did I seek to turn him back to the good side. Omar was long gone, fully dedicated to his bogus jihadi dogma. No way he'd come back into American hands or admit he'd made a mistake by joining a terrorist group that now wanted to kill him. I didn't want to persuade him, either; his conversations were working just fine. The audience watching our discussions—that was my target. American boys thinking about joining al-Qaeda, Somali Shabaab members distrustful of the Western recruits in the ranks, and Arab terrorists, their leaders, and their popular supporters who were once keen to recruit an American and now reeling from his disclosures.

The CIA uses the acronym MICE, but I prefer CRIME. The letters describe the motivations and enticements an intelligence officer or law enforcement investigator uses to recruit an agent overseas or a street informant in America. They're the reasons why people turn, or flip, why they begin reporting on a group they once declared allegiance to or betray an ally on behalf of their foe. *C* stands for *compromise*. Compromised people can be coerced into doing something they might not normally consider. A criminal charge, an outstanding arrest warrant, unpaid debts, a sick loved one who needs surgery—all provide avenues for convincing a person to provide assistance. I add *R* for *revenge*. Think Mandy Patinkin in *The Princess Bride*: "My name is Inigo Montoya. You killed my father. Prepare to die." There may not be any other motivation that makes people as relentless in its pursuit. The unjust murder of a loved one, wrongful treatment by others, perceived injustice by a rival: revenge, once pursued, usually can be countered only by death. *Ideology* constitutes the *I* and represents the purest motivation for any action. Those driven by ideology always prove the hardest to flip and most difficult to stop. *M* should be in bold: *money*. It's the most common reason for betrayal, and the flimsiest. Those incentivized by cash prove to be the easiest to recruit, and most likely to deceive or switch teams. Finally, the *E* is for *ego*. Fame and glory, the desire to be a hero, makes men do strange things. Empower and embrace the ego of a narcissist and he'll be a cost-effective asset, a turncoat for good or evil, depending on the suitor.

I went back to my notebook paper and sketched out a quick assessment of Omar Hammami's motivations for revealing so much on Twitter. The most obvious motivation for his endless disclosures was compromise. The more Omar got his story into the public regarding al-Shabaab hunting him, the more likely he'd be to survive, gain protectors, and push Shabaab into a no-win situation. If the terrorists killed Hammami, they'd hurt their brand in the eyes of future recruits and international supporters. Furthermore, each Shabaab attempt to hunt Hammami and quell his supporters increased Omar's revenge response. A war inside global jihad was what I wanted. I'd amplify any of Omar's resentment and accentuate his quest for revenge.

More subtle but still immediately apparent were Omar's egotistical motivations. Hammami loved attention—*loved* it. He thought of himself as a future jihadi visionary, consistently sought to showcase his theological expertise, and pined for the attention of senior jihadis and well-known counterterror experts. Omar wanted to be famous, and I'd help him do that. In so doing, I'd undermine motivations others might have for heading off to Somalia and joining al-Shabaab.

There were also topics I wanted to avoid when chatting with Omar. Hammami wanted to be an ideological expert, and he'd spent time studying and pontificating, developing his own vision for the future of global jihad. As a non-Muslim lacking any theological expertise, I risked empowering Omar by engaging in his religious rants and raising his profile among his supporters. I wouldn't be able to convince him that he was wrong about his religion, and I stood to look quite stupid if I tried and failed. A second area I sought to avoid was money, specifically his financial situation. He had left America to join terrorists in one of the most impoverished countries in the world. Sitting in prosperous America, I didn't want to glorify his financial sacrifice.

I took the three motivations I wanted to amplify and then identified common ground I could make with Hammami for rapport building. He wanted to talk to me—that was obvious—but I didn't want to speak with him strictly regarding terrorism. One heated debate would end our engagements. Persuading him to divulge more information or discuss his positions would mean first getting him to feel a deeper connection.

Omar and I had backgrounds that were similar in many respects. He grew up outside Mobile, Alabama, and I, having grown up a similar distance from St. Louis, Missouri, watched the same television he did, studied the same books in school, saw the same movies, and played some of the same sports. His longing for home increased his desire to chat with me, reach for my discussions, and seek my approval.

What made Omar appealing to Americans was that he walked and talked like an American, but this behavior also weakened his appeal among devout Islamists and jihadists. If he trashed al-Qaeda and al-Shabaab and joked it up with Americans on Twitter, he undermined his own appeal with Muslim communities whose affection he desired. Compromise, revenge, and ego—I would pursue these with Omar by playing to his American cultural tendencies. I'd write blog posts to amplify his disclosures and stroke his ego, then I'd sustain discussions through the nostalgia of America.

Anytime Omar drifted to American jargon, I'd slide with him.

"I c ur bikinis and raise u 4 wives in this life, 72 in next!" Omar tweeted in late January 2013, feeling strong with the increased following he had garnered on Twitter.

I'd follow shortly after with "come on Omar, don't tell me u did all this 4 the chicks, easier ways to meet women."

Mixed in between these lighthearted jabs would be debate over the future direction of jihad and terrorism. Hammami would read and critique. A tweet questioned why Hammami didn't support other jihads in Africa, and his response was eerily prescient:

"Syria is great. Many stages to pass b4 khilafah, and many ways things could go astray. But it'll work out, and they have tabuuleh!"

Omar was right. The Syrian civil war provided ample opportunity for young jihadis to achieve their dreams of state building and find the violence they so sought. Omar had opined on Syria and food, and there was my opportunity for common ground.

"Omar, question for you. Applebees, TGI Fridays or Chili's? You pick." I thought of some food options I imagined he might miss from Somalia. Omar's answer didn't disappoint.

"I'll answer 4 him: give me dat baby back, baby back, chiiliies *halal* baby back ribs, w/barbecue sauce."

After I fired American culture questions to Omar and he responded, I noticed that Omar's Twitter adversaries would begin tweeting in Arabic, pointing out to the global jihadi community that Omar was not like them—less devout, impure, and chatting with the enemy.

"Omar, lunch question. Buffalo wings, nachos or pizza? U pick," I posted.

Omar quickly replied, "hot wings."

Every week, I'd post a question or two on food. "Tonight for dinner Omar, what do u think? Chicken soft tacos, Caesar salad or Chinese take out (all halal of course)."

His response was as expected: "sucker 4 chinese, Didn't read the bio? General tsao, sweet sour, shrimp fried rice, baby corns n veg."

I imagined Omar drooling as he typed his response on his cell phone. The next time I passed a Chinese buffet in Boston, I snapped a quick picture. When I returned home, I uploaded the picture to my blog and blasted it to Omar: "halal BBQ style." He clicked on the picture, again giving me a Google Analytics signature for his mobile device, and followed with a response.

"Drool. Mmm," Omar replied to the tantalizing photo. "Gonna eat a chicken my wife slaughtered and cooked right now."

"U do that. I'll stick with the chicken wings" was my answer to Omar's post. Then he came back later with an evaluation of his wife's cooking: "Omg! It was old and free range! Rubber! The one last week was close to the real thing though."

I sustained this American banter with Omar, making him nostalgic about his childhood back in the States. "Omar, this morning's breakfast challenge, u pick Dunkin Donuts or homemade Biscuit with raspberry jelly?" Because I'd served in the Army in the South, I figured the question might entice him.

"That's a tuffy. Can I call a friend?" he asked.

Food discussions routinely brought Omar to American references from television, movies, and music. "Best thing since 'Jason.' crank up the chainsaw," he cracked one winter day, using a *Friday the 13th* movie character to justify terrorist violence. In other discussions, Omar reminisced about the kids' educational programming *Reading Rainbow* and *Wishbone*, where famous stories were retold through an adorable Jack Russell terrier.

The food dialogue discredited Omar among some of the terrorist audiences, but, more important, it opened a gateway for further engagement more in line with my goals. Within a month, jihadis in the ranks and even Hammami's fellow defectors were trying to silence him on Twitter.

"Funny how people contact me to say stop exposing the opp. Bcuz it make kuffar happy, but so few call 4 an end 2 the opp itself."

Quickly after, an e-jihadi shot back to this tweet, "Abu Mansuur [Hammami], tell us what you're after. Are you pro Jihad or Pro-Americans. we are doubting [y]our intentions."

Omar's discussions provided essential information about where he and his opponents resided in Somalia, creating a nice terrorist political map. Shabaab's supporters and their detractors tweeted about their surroundings. Over time, I could piece together their general locations, sometimes the exact towns from which they tweeted. I could discern which ones were true terrorists in Shabaab territory, which were Shabaab dissenters from lesser clans, and which were international diaspora members in far-off places like London, Stockholm, or even America.

To exacerbate these divisions and highlight the differences between the camps, I honed my blog posts to further entice Omar into engagement. Again I fell back on some old training. Using "RPMs" is a discussion technique used to nudge guilty people toward a confession. The science behind the approach remains a bit dubious, but even in general conversation, RPMs help bring the interviewer closer to the subject, making the person more comfortable divulging secrets or confessing to crimes. *R* stands for *rationalize*, and throughout my blog posts narrating Hammami's adventures, I'd note that Omar had "no choice" but to run from Shabaab and "do the right thing" ideologically. When not rationalizing his choices, I focused on the *P*— *projection*. "Shabaab betrayed you, they forced you into this position," I'd say, offering a convenient vehicle to saddle up to Omar's position and take his side. I was patting him on the back and confirming his actions as the smart decision, placing blame on Shabaab and noting that he had no choice in his decision. *M* is where you *minimize* the person's actions, and this proved a bit harder to use in Omar's case. He was a terrorist and he had killed people. So I'd drop in comments regarding how egregious Shabaab's atrocities were but then note that what Omar had done paled in comparison.

In addition to using RPM interrogation techniques, I'd play hot and cold with him. I'd lead one day, patting Omar on the back for being a fighter instead of one of these online wannabees chatting at us on Twitter. A day or two later, I'd criticize him for being so silly as to join a terrorist group that now wanted to kill him. Beyond that, I'd create false competitions. When Omar landed on the FBI's Most Wanted list, I noted that he was worth only a percentage of the top-valued American terrorist on the list, Adam Gadahn, a close adviser to al-Qaeda's chief, Ayman al-Zawahiri. Omar took it in stride, though, loving the attention and cleverly responding, "a friend in shabab jail was amazed at me being wanted by u.s. AND muj. Someone responded: like qadaafi!"

Omar, in turn, used his Twitter following, his personal metrics, to compete with me.

"Thnx 2 new followers. I generally like to engage those who take time 2 tweet me, but this is just ridiculous!" Omar said, congratulating himself for his growing Western fame.

"Clint, if I get another 300 followers or so, we'll be tied. Up for a private match of twitter fightin'?" Omar kept a close eye on his followers and mine, and now he wanted a personal chat. *No, thanks, Omar, I'll keep you wanting, and here talking in public.* Omar, isolated in Somalia's wilderness, was getting attached to his conversations with me and the other Westerners he'd left behind when he ran off to become a terrorist. The time had come to move conversations past rapport building.

Two months into our conversations, my relationship with Omar had reached the point where I felt I could challenge him with questions of substance. I wanted Omar to disavow al-Qaeda. I knew he wouldn't completely defame them, but I hoped he would suggest that the terror group's time had passed. Such an admission from a well-known American foreign fighter would take al-Qaeda's glamour down a notch. To get there, I focused on the conflict Omar had with one Shabaab booster. Jawshan, a Shabaab-supporting Twitter troll, clearly gave Omar pause, and I saw an opening.

"Instead of a twitter fight, could u & Jawshan set up a shabaab celebrity boxing match or something & broadcast it?" I posed the question to Omar, and he again thought he'd be clever.

"Somalis don't box, so I'm sure I'd win, but jawshan is more the hide behind the curtain and pretend 2 b wiz of oz type." Omar pursued the question as a way to tread on his competitor. A perfect setup for me, to side with his adversary and see if he'd reach for my approval.

"U complain about #sockpuppet Americans but then u argue with Shabaab #sockpuppet—they must get to u." I was offering some credibility to Omar's rival Jawshan.

"Shabab one is an arm pretending 2 b its own sep. body. The American one has a hand up its . . . sleeve." Omar's response fell flat.

"Took u while to think up this right. Nice try but ur attempt to be clever didn't work out this time . . . ur analolgy no good." I wanted to sting him a bit; he had gotten too comfortable with me.

"No. It was immediate, but net is bad. You can have ur opinion abt my analogies i guess. It's not what makes u a watty kafir!" Omar tersely replied.

"Ahh name calling . . . I guess when u don't have answers just fall back to that . . . like team jawshan." I wanted to keep up the pressure on Omar a bit.

"So someone banned big watty small's cigarette break at work and now he's taking it out on the terrorist." Omar appeared upset with me, and again he resorted to a negative nickname he and his small terrorist posse had assigned me weeks before: "Big Watty" or "Watty Kafir," a transliteration in Arabic of my last name, an insult roughly meaning "lowly unbeliever."

Now was the time to challenge Omar's decision to join a terrorist group. "I can't figure out why u insist on violence to pursue ur beliefs. U could have stayed in states & worshipped the way u want."

Omar's response was lackluster, scripted, and vacant. "Establishing khilafah is the pinnacle of worship." A *khalifah*—caliph, in English—is the civil and religious leader of a Muslim state and the representative of the Muslim god, Allah, on earth. Omar seemed to suggest that pursuit of violence must be done to establish this *khalifah*. The answer was again weak, and he knew it.

"So like I said, Jawshan gets to u then? Why do u care what he thinks? He should be worried bout u right?"

Strangely, my opportunity to prompt the question came during a workout. On an April afternoon in 2013, I glanced down at Twitter while I was at a gym along Commonwealth Avenue in Boston. Surrounded by Boston University students hopping on treadmills and staring at themselves in the mirror while lifting dumbbells, I saw the opening I'd been looking for.

Omar had tweeted, "only respond to things that could lead to doubts if led unanswered."

I paused for a second, then started typing on my iPhone.

"Ok, we'll, here is what I have doubts about when it comes 2 u, 'are u a member of al-Qaeda & do u believe in their ideology?'"

I challenged Omar's allegiance to al-Qaeda. I wanted him to disavow the group he'd once defended back in Alabama and aspired to on social media. I hoped he'd point out, in front of all jihadi watchers, that al-Qaeda's time was passing.

Omar answered quickly: "not member of aq but agree with most of their ideology." *Thanks, Omar.* Guys like Omar were still motivated to commit terrorism but looking for a new direction.

"Hey, man, are you using this bench?" As I tried to pull dialogue from a terrorist half a world away, a young BU student was getting frustrated with me. *Hey, man, I'm busy talking to a terrorist, give me a second.* In my mind, I was shouting at this kid, but he was right: I wasn't lifting, and my towel secured a bench I hadn't used for at least ten minutes.

"It's all you," I said, turning back to my Twitter feed, trying not to be that annoying prick everyone runs into at the gym who hogs the equipment while gazing at his phone.

Omar admitted that he wasn't a member of al-Qaeda, and again pointed to where jihad was headed: a decentralized stew of angry young men looking for a new direction and the next objective.

In the social media world, Omar was becoming more of a friend to the counterterrorists than to the terrorists. The more Westerners engaged him, the more information he coughed up, spilling the beans on infighting among jihadis in Somalia. As March turned to April, I played more and more to Omar's ego. Every disclosure Omar made I'd summarize and post on my blog. Any evidence emerging publicly of Omar's disclosures and reports of internal terrorist conflicts I'd add in there, too. A couple of days after our al-Qaeda discussion, I sat among BU students again, this time in my favorite coffee shop, Blue State Coffee, and quickly wrote a post supporting Hammami's thesis. "#Shabaab in #Somalia publicly fractures & pressures al Qaeda" went live just as I headed home for dinner.

I checked my phone as I walked, and there was Omar: "5 min ago was gonna ask u if were busy writting a new blog post." *You bet, Omar, I'm always there to help you hurt a terrorist group.*

A few days later, Omar provided an English version of how al-Shabaab had betrayed and hunted him in Somalia; he'd learned that most people in the Arab world weren't too interested in the tragic story of an American jihadi. I again repeated my technique, quickly summarizing his tragedy on my blog: "#Shabaab's betrayal of Omar Hammami—in English this time!"

"If u get some award for ur coverage, u owe me a halal burger." Omar loved reading about himself.

I joked back with Omar: "I haven't heard Pulitzer knocking, and my English is about like ur Arabic, so critics often overlook me." Again I lumped him and myself together, like long-lost Twitter brothers.

A couple of months into Omar's discussions with Westerners on Twitter, he rightly became worried about his personal security. Al-Shabaab, once the champion of Twitter, now sought control over its group members and their tweets. They understood what I and others were seeing and what we were doing—tracking the strength, unity, and direction of terrorist groups and their members using open-source social media information. Shabaab members scolded one another in tweets, referencing their boss's instructions to stop revealing operational information on Twitter. They also warned Omar about his disclosures.

"The only reason thy r even talking to u is to get more info frm you! And to make fun u dnt u see tht?" tweeted by one of Omar's Twitter followers.

I'm confident Omar knew we were tracking his information, but the message struck a chord with him. "Is this true?" Hammami tweeted out to his Western followers. Some immediately responded with denials, and others sought to redirect his discussions to other topics. But this represented an important moment in my relationship with Omar. Most will try to hide from the hard questions, obfuscate, or lie. But these misleading statements end relationships, both real and virtual. The best answers are simple, straightforward, and true. My response was easy:

"For information, yes, of course, I write blog posts about it, if u don't talk about something interesting, I don't right [*sic*]," I said, mixing up homonyms amid my daily fatigue.

Omar, in response, seemed all right with it. "Yeah, I thought it was a fun dramatic question to ask. I'm a grown up and know who I'm dealing w/. Good answers tho."

I then offered, "funny, u & I r 2 only people I see in these discussions that r actually who we say we r on this thing called Twitter."

Omar, a typical American, confirmed, "I like to keep it real yo. Word."

Our game of operational-security chicken needed to end, though. By tweeting with him, I was revealing a good deal about my own locations and activities and making myself vulnerable to whatever young jihadi boy might decide to knock on my door and behead me.

In Omar's case, he should have been worried, as even his most innocuous revelations provided details of his whereabouts. "Had some ice today. Gave the donkey a day off to enjoy the rain. The shabab are bluffing. No worries." But Hammami, in showing off his fearlessness, gave me and the rest of the world key details. His technical signatures hitting my blog didn't provide much other than his presence in Somalia, but his words gave me important context.

In just a few short Google searches, I pulled up maps from aid efforts in Somalia, showing roads and villages not typically displayed or easily correlated in Google Earth. A few minutes later, I printed a Somali clan map created by a college professor, showing the group Hammami said had helped him hide in Somalia. It doesn't rain much in that country, so I pulled the weather reports for the clan area, and it had rained in only a couple of places that day. Omar mentioned ice, a rare commodity in the Horn of Africa and one requiring electricity, so I looked for places within a thirty-to forty-five-minute donkey cart ride—five kilometers or so—from towns where it had rained that day. Then I looked back to the tweets where Hammami had noted people speaking out on his behalf or in line with his views. His supporters consistently mentioned only three or four towns, and all surrounded the areas witnessing rain that day. In less than an hour, using a home computer, I had two to three guesses as to where Hammami was hiding in interior Somalia. If I could figure this out from my house, Omar had reason to be wary of what others could do with his information—terrorists or counterterrorists, both of whom wanted him dead.

But Omar loved the attention and felt that, through his notoriety, he was waging a counter-jihad against his previous terrorist overlords. He kept dumping insights onto the internet, and at times I couldn't even keep up with his disclosures. More English versions of his previously published Arabic documents came out each week, and I also had to focus on my day job to pay the bills working as a cybersecurity consultant. Omar wanted my attention, wanted me to keep writing. Two days after the Boston Marathon bombing, he reached out to me wondering why I hadn't been quickly writing up a new post about him. "Clint is either saving the day in boston, or busy digesting my 20page or so doc . . . and maybe a chillidog or something." He was right: I was distracted. Living less than two miles from the end of the Boston Marathon route, I'd been busy keeping an eye on the rapidly unfolding events.

Omar was a lost soul, estranged from his country, his family, and even his terrorist group. Hunkered down in a nearly inhospitable African country, hunted by his former fellow terrorists, he stayed alive for a time because of his social media prowess. Shabaab, which had pioneered Twitter use for terrorism, now suffered from the openness it had unleashed. To silence dissenters as the group declined, Shabaab turned off cell towers, squashing social media usage and information sharing among its challengers. When Shabaab raided Omar's home, searching his family's belongings, they took the cell phone charger, a scarce commodity in Somalia and one that silences Shabaab's mobile-phone-empowered critics. Words, information, and social media had become more powerful weapons in Somalia than AK-47s and RPGs.

By the end of April 2013, Ahmed Godane, leader of al-Shabaab, had had enough of Omar's disclosures and the dissension they had fueled, both in his ranks and among clans in Shabaab's orbit. Not long after complaining about not having a proper shower since he arrived in Somalia, Omar tweeted, "just been shot in neck by shabab assassin. not critical yet."

Soon after, he posted a picture showing his neck wound. Omar dropped off Twitter for a few days, then resurfaced to detail his tribulations leading up to being wounded. He'd appealed to a Shariah court for help, and while the local court took Omar's side, Shabaab's hunters didn't stop. Omar and his small posse headed into the woods, where a small battle ensued. Omar survived, and Shabaab's troops took losses. But Omar's social media presence would never be the same. He tweeted one last, hopeful time on May 3, 2013, saying that forces were closing in on him and that the people were on his side. And then silence: Omar vanished from Twitter, much as he'd done for extended periods before.

The summer passed, and I moved on to other terrorist groups, namely the brewing battles between jihadis in Syria. No word came from the @abumamerican Twitter account for months. Then, just after Labor Day 2013, Omar briefly reappeared. Harun Maruf, from Voice of America, did a phone interview with him, and excerpts appeared in a story on September 5. When asked whether he'd come back to the United States, he said, "That is not an option unless it's in a body bag."[6] He whined in one tweet about the interview, claiming that Voice of America hadn't provided enough of his thoughts and had simply painted him as a terrorist. September 5, 2013, would be Omar's last tweet.

Omar Hammami didn't come home to the United States, in a body bag or otherwise. Twelve years and a day after the September 11 terrorist attacks, al-Shabaab hunted him down in the forests of Somalia and killed him.

News reports surfaced as word spread quickly. A Hammami ally took control of Omar's Twitter account, sending out a death announcement to the world and to Omar's father, Shafik Hammami: "We confirm the martyrdom of Omar Hammami in the morning of Thur 12 2013. Shafik's family please accept our condolences." Omar's discussions with his father via social media offered a twisted, complicated subtext to his writing. Shafik still lived in America, and he slowly watched terrorists hunt down his terrorist son, all the while concerned for his Somali grandchildren he'd never met. For a few weeks after Omar's death, the @abumamerican account continued to promote Omar's story and the failures of al-Shabaab, before falling silent.

Harun Maruf released his full telephone interview with Omar on the day of his death. The broadcast eerily outlined the trap Omar was caught in. Hunted by both terrorists and counterterrorists, hiding in the woods with a handful of supporters, Hammami could never leave Somalia, nor could he remain there alive. Until the end, he envisioned himself to be a pseudo-cleric, explaining, in a slightly southern American drawl, why those who didn't follow his brand of Islam were wrong and could be subjected to violence.

"You're either pregnant or not pregnant, you're either are or you're not . . . you need to follow all of the Shariah or you are not a Muslim." Alabama analogies mixed with jihadi elitism—only the bizarre world of social media could create such a collision of cultures.

Omar and I weren't friends, because he was a terrorist. As an Army officer or an FBI agent, if assigned the mission, I would have hunted him down as directed, arrested him if mandated, killed him if necessary, and not thought much about it. Even as I watched his slow demise via Twitter, I didn't feel a bit of surprise or remorse when Shabaab killed him. Omar started digging the shallow grave he ended up in the day he started backing al-Qaeda's violent ideology. He sealed his own fate when he defected from al-Shabaab and sought his own fame and notoriety on Twitter. If it had been up to me, I would have liked to see the Somali clan Omar had gone to for protection turn him over to the United States for a $5 million reward. I had hoped that Omar would eventually stand trial or make a deal with his native country, becoming a useful defection from terrorists' ranks and counter to jihad's bankrupt dogma.

When I communicated with Omar about America, he sounded like so many soldiers I had met in the Army. If he hadn't become a terrorist and gone off to kill people, if we'd met at another juncture back in the States, I expect we'd probably have become friends, sharing a box of Krispy Kreme doughnut holes, laughing about a silly eighties movie we watched as kids, or shooting guns at coffee cans in a cornfield or a quarry. The weirdest thing I've realized in the global war on terror is that very little differentiates the average foot soldiers on each side of the war. Terrorists and the counterterrorists who chase them have more in common with each other than the al-Qaeda propagandists and American politicians who give them marching orders. All that separates them is a decision: not the decision to kill or not to kill, but the one about why they do it—militant Islam or democracy, freedom or Sharia law, money or piety, family or faith.

Omar Hammami, trapped amid a failed Shabaab experiment to create a state, anticipated not only the move of young jihadi boys to Syria, but the nature by which al-Qaeda would unravel. While American counterterrorism analysts pontificated about an al-Qaeda comeback, Omar noted, in January 2013, what was to come: "terrorism is a staple so long as shari'ah is followed. Post-aq = broadbased jihad."

The Islamic State harnessed this notion, the movement of young men to social media, untethered from the direct orders of jihad's forefathers, bin Laden and Zawahiri. These terrorist leaders advanced violent jihad's goals and direction but stymied the zealousness of the "me" generation rallying on social media. Months after predicting a post-al-Qaeda era, Omar Hammami again correctly sensed the global direction of jihad, noting, "We [jihadis] were previously too focused on spreading 'jihad' globally, now on establishing 'shari'ah' locally. it will equal out to become khilafah." Omar died before ISIS became the Islamic State, but he could see it coming. As Hammami and his fellow defectors died in Somalia, jihad's next generation launched a civil war in al-Qaeda's ranks in Syria.

I closed out my time with Omar looking at a family photo he'd posted on Twitter. Omar, rifle in one hand, knelt down with his other hand on his wife's shoulder. His wife and daughter, draped in full burqa, sat next to him in front of a makeshift tent. America meets jihad. It was the equivalent of an Alabama family picnic photo in a Sharia world, and it didn't work. The two women could be anyone, fully covered and completely unrecognizable. Why put them in the picture? Never, in my years of scanning terrorist pictures on social media timelines, had I observed a devout jihadi taking family photos with fully covered women. Omar was proud of his family, but it didn't fit with the new culture he'd adopted. His story, like those of so many Western terrorists, was confusing and conflicted. I wonder whether, in his final moments in Somalia, as assassins closed in to kill him, he was still certain in his ideology, or in doubt. Did his dying thoughts reach for jihad and martyrdom, or his home in Alabama?

As for Shabaab in Somalia, after killing Omar Hammami and executing the spectacular Westgate Shopping Mall attack in Kenya less than two weeks later, it continued its steady decline. Defectors from its ranks grew, and just less than one year after Omar's murder, a warhead hit the forehead of Ahmed Godane. A U.S. drone strike had killed Shabaab's murderous leader. Today, Shabaab hasn't vanished, nor will it. One jihadi terrorist group or another has called Somalia home for decades, and some violent fringe will remain through our lifetimes. But Twitter, the platform the group had used to reach new heights, ultimately helped unravel them from the inside out.

Rise of the Trolls

"Who are these assholes?" I whispered to myself.

By 2014, social media was becoming more antisocial by the day. I stared at my iPhone's Twitter feed on a cold January morning, watching a fresh batch of anonymous haters.

I first came to the Twitter platform as a promotional tool for SelectedWisdom.com. Many of the top hands in counterterrorism and foreign policy circles were using it, and commentary and discussions there often proved quite useful as a form of debate. But that changed at some point. Every day Twitter was increasingly a space where the angry, the disgruntled, and the unhappy unleashed their rage on the world. I had been trolled before, but these trolls, emerging about a year after my discussions with Omar Hammami, were relentless. Something was different about them.

A few days before the troll storm dominated my Twitter feed, Michael Doran, Will McCants, and I published an article in *Foreign Affairs* website challenging conventional wisdom regarding negotiation with those who may harbor al-Qaeda sympathies. "The Good and Bad of Ahrar al-Sham" offered that the United States might negotiate with an Islamist group to challenge the influence of al-Qaeda's growing Syrian affiliate Jabhat al-Nusra. Americans loathed the thought of many troops returning to the Middle East, and we posited that this soft-power approach might be worth exploring as an indirect route to curbing terrorist growth amid Syria's civil war. The trolls didn't agree.

These Twitter trolls were different from most trolls I encountered, though. Nearly all labeled me a terrorist supporter and sympathizer. Regardless of where they proclaimed residence—Europe, North America, Asia, or Australia—all agreed that Syrian president Bashar al-Assad's heavy-handed rule provided the only deterrent to terrorists overtaking the Levant. The trolls kept it up for days, then weeks and ultimately months. Each morning I'd wake to fresh hate from the same accounts, which would be supported by nearly unending waves of retweets at roughly matching intervals. Dozens, sometimes hundreds of odd-looking accounts would favorite some of my tweets. I wasn't the only one who noticed this phenomenon. Two colleagues from my counterterrorism days noticed the trolls as well, and they both had more technical know-how than I did regarding social media analytics.

Andrew Weisburd was a natural social media savant. He could look at an online persona, spot it as friend or foe, and trace its connections to a host of bad actors online. As a hobby during the 2000s, Weisburd began tracking al-Qaeda online from his couch. He identified and outed terrorists lurking on the internet so well that al-Qaeda fanatics mailed a white powder package to his house, along with a death threat:

"To the Jewish asshole Aaron [sic] Weisburd. This is our donation to you. Either you close the website called Internet Haganah by next week or you will [be] beheaded."

An FBI inspection found no signs of anthrax, and Andrew resumed his hobby with a bit more caution. Not long after, his hobby became his work, and he eventually joined me at the Combating Terrorism Center at West Point. Andrew helped train law enforcement, the military, and academics on how terrorists actually operated on the web—debunking much of the media hysteria present at the time.

By 2014, Weisburd had been doing his work for a decade, and

through the course of endless investigations he'd identified thousands of nefarious accounts, many of which surfaced among the troll armies messing with me and others studying the Syrian conflict.

Weisburd connected some of the trolls to a recent internet nemesis: the Syrian Electronic Army. The SEA presented itself as a new hybrid threat of the online world, embodying the spirit of more popular hacktivist collectives such as Anonymous and LulzSec, but clearly in the bag for President Assad and the Syrian regime. The SEA effectively became the first nation-state cyber proxy force on the internet.

The SEA had undertaken a string of attacks against American companies since 2011, hitting a wide range of targets but taking aim at many mainstream media outlets that were revealing the horrors of the Syrian civil war. The *New York Times, Forbes,* the *Sun, 60 Minutes, 48 Hours,* the *Sunday Times,* the *Independent, Time Out,* NBC, the *Daily Mail*, and *Le Monde* all suffered attacks, with varying consequences.

Other than a brief encounter of a few seconds, I'd never met J. M. Berger except on Twitter. After connecting on social media, he and I learned that we lived only a few blocks from each other in Boston. J. M. had a gift for social media and for terrorism research. His interviews and conversations with terrorists, notably Omar Hammami and those Shabaab supporters challenging him, had allegedly earned him the nickname among e-jihadis as the "Loud Shaytan"—a transliteration of "Satan"—a nod to his ability to sow doubt among terrorist conversations. J. M. noticed this new outbreak of trolls as well.

Most important, J. M. had built his own method to tag and track bad actors who didn't want to be seen on Twitter but still wanted to influence audiences of supporters. While Andrew, the "Jewish Asshole," dug into the trolls, J. M., the "Loud Shaytan," inhaled the social media storm surrounding the SEA, narrowing in on the key centers of influence among this pro-Assad swarm. I, "Watty Kafir," partook in some Twitter conversations with nefarious accounts and attempted to map out what the trolls were trying to do. The more we dug, the stranger things became.

On March 21, 2014, a petition was posted to the WhiteHouse.gov website. "Alaska Back to Russia" called for the United States to return its largest state back to the country from which it was purchased in 1867.[1] Petitions ranging from the benign to the insane populate the WhiteHouse.gov website all the time. This one wouldn't have been any stranger than most had it not received more than 39,000 online signatures within only a few days of its inception.

Analysis of the Twitter accounts posting on this petition showed an odd mathematical pattern. When the signatories to most petitions on the White House website are amassed, they follow a highly normal distribution. Some Twitter accounts have only a few followers but follow many accounts. Others have many followers but follow only a few. In between these poles lies a natural blend of accounts, with most in the sample having roughly similar numbers of accounts following them and accounts they follow.

The accounts posting the "Alaska Back to Russia" petition didn't follow the normal distribution pattern, though. Their followers/following combinations were incremental and mathematical, falling in repeated intervals, which suggested that they had been manufactured through the use of mathematical algorithms (i.e., social bots). Oddly, the bots hammering away at the "Alaska Back to Russia" petition overlapped with the pro-Assad trolls seen at the start of 2014 and the SEA hackers breaching American companies for several years. Even more curious were the SEA's connections to previously identified social media propagandists for Hezbollah and Iran. Strangest of all, though, were the accounts' linkages back to Russia. Many were tweeting in English and Russian and the trolling effort clearly and overtly sought to embarrass the White House.

Effective troll armies consist of three types of accounts: hecklers, honeypots, and hackers. Hecklers lead the propaganda army, winning audiences through their derisive banter and content-fueled feeds. Hecklers identify and drive wedge issues into their target audiences by talking up online allies and arming them with their preferred news—

both true and false information, loaded with opinion, that confirms audience member beliefs. Hecklers also target social media adversaries and focus the angst of their cultivated supporters against opposing messages and their messengers. In the case of Syria, for example, anyone pointing out President Assad's human rights violations might immediately be called a terrorist sympathizer and subjected to endless 140-character jeers and taunts. Oddly, examination of these hecklers' Twitter and Facebook accounts revealed few followers that seemed real, nor did they post pictures of themselves with friends or strolling in their local neighborhoods.

When hecklers alone can't stop the challenges of the opposition, honeypots sweep in to compromise adversaries. Honeypots, in the traditional espionage sense, are attractive women who seduce men into compromising sexual situations. On Twitter, females remain the predominant form, but they can also assume the persona of an allied political partisan. Among the SEA, attractive females—or what appear to be women—performed the traditional mission of befriending men in the target audience or sidling up to adversary accounts hoping to compromise their personas or publicly embarrass them. "Can you follow me so I can DM you something important?" might be the siren song for one of these e-ladies. Lady honeypots in 2014 were seeking follower relationships with men, which would lead to a direct message capability. Once virtual relationships were sealed, honeypots could deliver a link via a direct message, one that would allegedly lead to privileged insights but often contained a malware payload allowing them to gain entry to a target's computer. Not all honeypots were lovely ladies, however, and some posed as motivated political partisans mirroring the views and banter of their targets. Hyperpartisan liberals and conservatives would appear to befriend rabble-rousers along nearly any political pole worldwide.

Behind the scenes, but still observable in the SEA social media storm, were hacker accounts. Examination of their follower and following relationships showed that they were highly networked with honeypot accounts, likely controlling the conversations between the lovely lady personas and their unwitting targets. The malware that honeypots delivered to unsuspecting men opened doorways to their phones and computers, causing them to give up their personal emails, corporate communications, in some cases, and their contact lists, allowing for malicious spam distribution.

Honeypots and the hackers behind them waged a highly successful campaign across a swath of companies and Western personalities in 2013 and 2014. Corporate America suffered as unwitting employees clicked onto malicious links and in turn coughed up access to private databases of subscribers and workers. The most damaging of these breaches came in the late spring of 2013. After claiming responsibility for breaking into the *60 Minutes* and *48 Hours* television shows' Twitter accounts, the SEA's most damaging alleged hack came on Tuesday, April 23. The Associated Press Twitter account posted in the early afternoon:

Breaking: Two Explosions in the White House and Barack Obama is injured.[2]

The @AP Twitter account had more than 1.9 million followers at the time, and the SEA's false information sent the stock market plunging 1 percent in a matter of minutes—the equivalent of $136 billion in value—before recovering five minutes later. The Associated Press Twitter account returned to normal social media operations the next day, but the implications of the SEA's actions were clear: hacking to steal information was old hat; hacking to influence would be the future.[3]

The troll army's heckler accounts, those that began challenging me in 2014, could work with hackers and their sister honeypot accounts or they could operate independently of these social media allies. Their tactics were far more ingenious and insidious than those of even the most talented internet propagandists of al-Qaeda and the

Islamic State. The hecklers weren't hacking people's computers; they were hacking their minds, in two ways. In one sense, they sought to change a target audience's perception on issues, nudging audiences toward preferred foreign policy positions and influencing experts, politicians, and media personalities toward a pro-Assad or pro-Russia stance. When not shaping audience conversations through a barrage of slanted content and supporting banter, hecklers sought to batter adversaries off social media platforms through either endless harassment or compromise.

Over a three-year period, accounts, primarily on Twitter but also on LinkedIn, would attempt to befriend me. Rather than chide me for publicly being anti-Assad or anti-Russia, these accounts would attempt to send me helpful tips, like my old friend Omar Hammami and his cronies had. News articles that supported my academic think tank writing would be fired my way in hopes that I might follow the referrer's Twitter account. If I didn't bite on their support of my research, friendly honeypot accounts tried to play to my ego—heaping on praise for my military or FBI experience or even pulling at my heartstrings regarding an article I authored about my daughter's autism.

Whatever the avenue, once I followed the befriending Twitter account, I'd be peppered with direct messages. Conversations generally started with friendly banter about my research or an issue I'd publicly discussed on Twitter. Invariably, though, dialogue would switch to a range of issues, all of which sought to edge me to a pro-Assad, pro-Iran, or pro-Russia stance on foreign policy issues. On select occasions, an account might try to introduce me to other online personas or place me into group discussions of real people and social media personas, some of which I knew and some I'd not noticed before. These group conversations sought discussion among like minds and hoped to move my thinking to a foreign policy position through what appeared to be innocent dialogue. If I wrote an article challenging the authenticity of Saudi counterterrorism partnerships and methods, adversary accounts might point out in private chats how Assad—and, by extension, Iran and Russia—represented the only reasonable alternative to jihadists taking over Syria. They relentlessly pushed the same basic theme: *Who is better to govern Syria? A strong Assad or the terrorists of al-Qaeda and the Islamic State?*

Having trained at the FBI Academy and studied Soviet politics, military, and intelligence at West Point, I recognized the technique as the digital update to age-old spycraft. These personas were recruiting me, virtually, through the timeless approach of "spot, assess, develop, and recruit." Either my credentials or my conversations led these personas to spot me as an adversary needing to be compromised or a potential ally needing to be curried. By evaluating my public social media posts, they assessed whether I'd be amenable to connecting through a follower relationship allowing one-to-one communication. They then sought to develop the relationship through private direct message conversations. Sharing and discussing, these accounts hoped I'd move to their position, or, at a minimum, confirm the legitimacy of their stance. Finally, if the relationship developed as they hoped, the influence accounts intended to recruit me, wittingly or unwittingly, to promote their policy agenda in my own public messages or rebroadcast their messages in a public forum. Social media provided an avenue for these influencers to further cloak their true personas and real intentions, all with remarkably lower effort, less risk, and fewer resources than in the Cold War days of spy versus spy.

If I wasn't biting on a following relationship that permitted direct communications, they'd switch to publicly counter me rather than privately convert me. Countering my comments and those of other foreign policy pundits allowed them to pursue several parallel strategies. Through endless harassment, hecklers sought to make my and many other foreign policy experts' experience so negative that we'd remove ourselves from the social media platform altogether, take ourselves out of the debate, and create more space for the heckler's preferred foreign policy position. If we didn't leave the platform, we'd suffer endless challenges, with the goal of leaving us virtually wounded, our credibility in question.

A second, more involved countering technique was account compromise followed by public shaming. In this case, hackers, appearing as adoring fans, would bait targets into clicking on links, allowing them to access a target's computer and uncover their personal secrets or, in some cases, simply take over their social media persona. Account takeovers might result in the target's audience of supporters unfollowing the account for fear of being hacked themselves—something like public fears of communicable disease transmission. If the account takeover yielded particularly sultry and damaging secrets, the target's personal files would be selectively slipped to a media outlet, resulting in the target's possibly being fired from a job, hated by one's family, and certainly unfollowed and rebuked by even the most ardent online supporters.

Not surprisingly, during the summer of 2014, one thorny expert who'd been seen jabbing at the Kremlin had his private conversations of a sexual nature and a nude picture leaked to media outlets, a traditional technique deployed by online honeypots. The pundit immediately suffered embarrassment and investigation into potential misconduct based on his professional position. The public smear campaign worked for the moment, the professor took a break from both social media and work for a spell, and the trolls racked up another score. Did the Kremlin perpetrate this honeypot operation? Who knows. There were likely many people motivated to take down this expert, or maybe this was just another case of poor judgment, but the character assassination technique proved rampant in 2014.

I myself had to decide during these heckler conversations, whether openly adversarial or seemingly friendly, if the risks were worth the reward. Conventional wisdom said that I should immediately block hecklers yelling at me over Syria and honeypots wanting to chat about all things pro-Russia. One wrong click would bring hackers into my computer systems, where they'd have access to all my private information, which they'd surely manipulate to discredit me publicly. I could create parallel computer systems, one protecting my personal information and another for conducting social media research, but the costs and inconvenience made work nearly impossible.

I recalled the lessons from the Omar Hammami days: how much easier it is to understand terrorists if you talk to them rather than try to avoid them. I knew it would be far more difficult to figure out who these accounts were and what they were up to if I didn't stay engaged with them. Sure, I might go down in social media flames, but understanding an enemy requires one to engage with them, mess with them a little bit. I kept the direct message conversations open, I didn't click on their content, particularly links, and I'd only watch each day, praying that slipping my phone into my pocket or a bag wouldn't accidentally cause it to click on a nefarious link. I started taking some countermeasures as well, creating misinformation on my email accounts, posting pictures misdirecting my location, and leaving some stale digital bread crumbs on my system.

Most social media analysis in 2014 focused almost exclusively on the Islamic State, and that was where I initially encountered the Kremlin's trolls and their Iranian and Syrian allies. The Islamic State's social media radicalization and recruitment methods were unprecedented in the terrorism world, but they paled in comparison with what was emerging from Russia. The Islamic State mastered online shock-and-awe footage, and its coordinated dissemination to followers across a wide range of social media platforms proved unprecedented for extremists. But the hacker-honeypot-heckler collective of the Kremlin coordinated seamlessly in their targeted trolling, and they sought not to get young men to travel to Syria but to change international minds about the conflict overall—creating doubt over Syrian human rights violations, offering Assad as the only option to create a safe and secure Middle East absent of terrorism. Recruiting a few thousand disenfranchised boys into a terrorist militia had been done twice before—the Islamic State did it the best—so the effect wasn't new. But influencing the entire world, unwittingly, to accept President Assad and his indiscriminate killing and destruction of his own country—that was an unprecedented social media undertaking. The troll army we encountered in 2014, the one hammering away at the "Alaska Back to Russia" petition, were on the cusp of a new era of digital influence—they were employing automation alongside humans, creating a new, alternative, and false perception of the world. This was master-level influence unlike any the world had seen.

\\\\\\\\\\\\\\\\\\\\\\\\\\\\\\\\\\\\\

Computational propaganda leverages social media to rapidly spread information supportive of a particular ideology. One of the first groups to study this emerging field of online influence was the Oxford Internet Institute. Phil Howard, professor and leader of its Computational Propaganda Project, defines computational propaganda as "the use of information and communication technologies to manipulate perceptions, affect cognition, and influence behavior." This manipulation occurs through the deployment of what are known as social bots—programs, defined by a computer algorithm, that produce personas and content on social media applications that replicate a real human. These social bots have also passed the important milestone known as the Turing test, a challenge developed by Alan Turing, the great member of the British team that cracked the German Enigma code.[4] The test assesses whether a machine has the ability to communicate, via text only, at a level equivalent to that of a real person, such that a computer—or, in the modern case, an artificially generated social media account—cannot be distinguished from a live person.

The bots we observed did just that: they created artificial accounts, emulating real people, that mimicked the conversations of target audiences in several geographies around the world. Some bots were strictly automated spamming accounts that replicated the same message or a combination of messages at a standard interval. Other bots operated more as cyborgs, part human and part automation, such that an individual user would administer what may be dozens of replicated accounts, amplifying messages and selectively replying or messaging key influencers in an audience space.

Theoretically, bots could be employed for positive purposes, such as public awareness and emergency notifications. But Howard points out the dangers of social bots, noting that political actors in democracies might use them to promote political campaigns, transnational actors could easily employ them to influence foreign audiences, and authoritarians could leverage them for domestic social and political control.[5] Beyond rapidly spread falsehoods throughout specified audiences, bot amplification of news stories can lead to the widespread dissemination of misinformation among mainstream media outlets, overwhelming the ability of journalists and fact-checkers to assess the veracity of an endless stream of propaganda. Clayton Davis, a researcher of computational propaganda at Indiana University, notes how bots can create a "majority illusion, where many people appear to believe something . . . which makes that thing more credible."[6] Users must then evaluate an endless series of tweets and posts repetitively, evaluating the veracity of messages and their sources.

Our analysis of the troll army continued throughout 2014. Regardless of the foreign policy issue, troll army accounts deliberately pushed the same pro-Russian content to their targeted audience. News stories from Russian government-run outlets, along with a range of content from lesser-known conspiratorial websites, surfaced nearly simultaneously from social media personas that appeared, on the surface, to come from different geographic locations around the world. The networks looked less Syrian and more Russian every day.

The troll armies I'd encountered gained steam through 2014 and early 2015, and their patterns became predictable and repetitive, with one interesting twist: they were more interested in American audiences. American- and European-looking accounts among the troll army gained momentum, increasing their organic following among real Americans. Heckler accounts pounced on any issue, liberal or conservative, always taking a divisive tack—fomenting divides between rich and poor, black and white, immigrants and non-immigrants, and all those harboring antigovernment angst. If a Black Lives Matter protest broke out in a U.S. city, the trolls were there, pushing a mix of true and false messages with a conspiratorial bent. Remnants of the Occupy movement or conservative gun owners—it didn't matter to the trolls; they repurposed existing inflammatory messages or helped seed new conspiracies tearing down the U.S. government. The trolls gravitated to Russian state-sponsored outlets, and helped distribute their false news stories or manipulated truths to newly won Western audiences. Americans of all kinds lapped up this content, but the right-wing fringe groups ran with the content far more than the rest, particularly white supremacists and antigovernment groups. Still, for the first eighteen months or so, I didn't believe the Russian influence efforts on America made much difference. That is, until the summer of 2015.

One campaign promoted by this Russian troll network resonated more than others. Jade Helm 15 was a military exercise sponsored by U.S. Special Operations Command, held in seven southwestern states. The training exercise sought to improve coordination and operations among U.S. special operations forces, ostensibly to prepare them for overseas deployments combating terrorists.

Right-wing conspiracy theorists thought differently about Jade Helm 15. Bloggers, Alex Jones of *Infowars*, celebrity columnist Chuck Norris, and even Texas governor Greg Abbott all feared that the military training exercise constituted a secret U.S. government plan to "impose martial law, take away people's guns, arrest political undesirables, launch an Obama-led hostile takeover of red-state Texas, or do some combination thereof," the *New York Times* reported.[7] Russian troll networks saw fresh opportunity to further foment this division between the American public and their government. Hecklers infiltrating right-wing audiences shared and recycled conspiracy theories with unwitting Americans helping fan antigovernment flames. Russian government-run news outlets jumped into the fray, further promoting conspiracies of a Texas takeover.[8] Several hundred Texas residents showed up to protest and shout down a U.S. Army officer providing a public briefing on the exercise. The conspiracy reached such heights that Governor Abbott deployed the Texas National Guard to monitor Operation Jade Helm and ensure that no federal takeover was afoot.[9] Ultimately, the exercise occurred with little fanfare and was more of a public relations crisis than a threat of martial law. But the Jade Helm 15 exercise revealed to me and likely to the Russians just how easy social media influence could be with segments of the American public.

The troll army's operations in 2014 and 2015 indicated how Russia's long-run investment in state-sponsored media outlets was beginning to pay big dividends. Social media provided a subtler pathway for government-sanctioned news outlets like RT and Sputnik News to grow their American viewership. RT had launched in late 2005 as a satellite news channel under the moniker Russia Today. Viewership lagged under the overt Russian banner; its content had limited reach with foreign audiences. In 2012, Russia Today deftly changed direction, shifting its focus from Russian domestic issues to Western social problems and the flaws of democracy. Programming shifts coincided with a branding switch and the name Russia Today was condensed to RT, subtly masking the channel's sponsor.

In 2012, RT showed the fastest growth among international news channels in the United States. It implemented a social media strategy to spread among English-speaking audiences, becoming an early adopter and highly effective disseminator on YouTube. Today, RT sustains nearly four times as many YouTube subscribers as CNN. RT's Facebook chatter outperformed that of BBC World. RT achieved this growth, in part, by using its reporters' and producers' social media personas as distribution mechanisms for content that does not appear directly on its television channel, further obfuscating the Russian source. Disenfranchised Americans angered at perceived social grievances or bitter with the U.S. government, unaware that RT was sponsored by Russia, devoured this social media content, which confirmed their a priori assumptions, regardless of whether the content were true or not.

Throughout 2015 and leading into the election year, friends shared RT news stories with me on Facebook for the first time. Sometimes I'd point out to the sender that RT was a Russian state-sponsored news outlet. This usually caught the sender off guard; they'd be surprised to find out that the English articles they were reading and sharing came from a Kremlin outlet. When jumping in a New York City cab or traveling on the subway, I began noticing RT advertisements for the first time—ads promoting notable U.S. personalities. For instance, RT's Election Night coverage for the 2016 presidential election starred Larry King, formerly of CNN, Ed Schultz, formerly a left-leaning host on MSNBC, and former Minnesota governor Jesse Ventura, one of the world's most notable conspiracy theorists, who regularly claims that the 9/11 attacks were the work of the U.S. government. This lineup points to the range of audiences Russia seeks to influence—left, right, Occupy, white supremacist, antigovernment, those frightened by vaccine conspiracies, and climate change advocates *and* deniers. RT has something for every American if it means weakening support for the U.S. government and its institutions. RT's tagline is "Question More." Of course, it seeks to provide not answers, but doubt. *How do you know? Could it be this instead? Can you really trust the government? Isn't the U.S. government hypocritical?* The net takeaway of RT coverage is that nothing can be trusted, and if you can't trust anyone, then you'll believe anything.

Social media provides the perfect avenue for disseminating content to those who want a certain type of news but don't necessarily want to know how their news is made. Russia intelligently uses a range of nefarious state-sponsored outlets to promote its preferred narratives and quench the thirst of the conspiratorial. Russia's second-most-prolific outlet is Sputnik International, a wire service and online news site, which offers a more sinister and spectacular version of RT's mainstream coverage. If a claim against America seems a bit too outlandish for RT, which attempts to maintain a higher degree of credibility, Sputnik quickly steps in to proliferate the sensational. Beyond Sputnik, Russia foments chaos with the support of fringe websites. Conspiratorial sites from Moscow to America repeated Kremlin talking points and manufactured conspiracies throughout 2014 and 2015. Sensational stories built from Kremlin-conceived conspiracies pushed clickbait headlines into American Twitter feeds and Facebook News Feed bubbles. The more Americans clicked on these unknown websites, the more their stature rose in the mainstream media.

Russia's success with overt state-sponsored news and covert trolling can be credited partly to the seamless integration of its intelligence and security services. Each day, Putin's top deputy attends RT's daily meeting, helping direct the outlet's themes and objectives. Some Kremlin-linked trolls receive guidance from Russian intelligence services, but others are more loosely connected. Some die-hard supporters inhale Kremlin propaganda and belch it back out into social media at what seems like an unimaginable pace. Others are likely unknowingly being manipulated. For instance, a tweet or post about a conspiracy theory, political scandal, or national emergency might suddenly receive thousands of likes, retweets, and shares, incentivizing the original user to post more content and reap social media rewards. The user may think he's quite popular and never truly know that Russia was directing engagement at him.

But some—the core of the troll armies, like the Internet Research Agency cited in the Mueller indictment in February 2018—clearly receive marching orders from the top of the Russian government. Housed in a seemingly corporate building is an office for Kremlin-employed trolls who drive pro-Russian themes in chatrooms, blogs, comment columns, and social media posts. Trolls have a quota they must hit each day, just as in any other job.[10] Some of their articles, posts, and jabs are pithy and on target. Others appear crude and lazy, attempts to meet a deadline more than an objective.

Up until 2015, I had been trolled, but not hunted by the trolls. That summer, I traveled overseas to consult for a private-sector client. On the road with a couple of colleagues, I let them operate my dating app account as we made our way to Asia. We had joked while waiting at the airport about one profile, an overtly beautiful Russian woman I likened to a "Red Sparrow"—a female KGB agent specially trained to seduce Kremlin targets into sexual trysts. Jason Matthews, a former CIA case officer, had recently penned a bestselling thriller of that title (now a feature film with Jennifer Lawrence), and I remarked as I boarded the plane how easy it must be for the Russian intelligence services to compromise targets in the social media era on dating apps like Tinder, Bumble, and Match. I matched with the "Red Sparrow" account and didn't think much more of it for the rest of the transcontinental flight—matching and messaging bizarre accounts came with the online-dating territory.

Three days into my overseas trip, I logged back into the dating app and noticed a message from the "Red Sparrow."

"Thank you for visiting former FBI Special Agent Clinton Watts. We hope you enjoyed your stay at hotel _____. Our agents fluffed the pillows for you."

They most certainly knew about my travel in the country. My dating profile didn't have my full name or all my professional details, nor did the limited conversation I'd had with "Red Sparrow" provide any clues about my patterns. Ascertaining this information wouldn't be difficult, though—a quick screen scrape of my profile picture and some Google searches would give one enough details to render my true identity. As for travel, my staying in a hotel was logical—they didn't need to actually break into my hotel room or even know where I stayed to intimidate me. They only needed to create the impression that they were spying on me to send my imagination running. A smart play all around, and, who knows, maybe they did break into my hotel room and plant listening devices or surveil me. But this would have been a giant waste of their time and resources, as they would have found out I was no longer a government man and instead pushing PowerPoint slides in corporate conference rooms.

The Russians also intelligently waited to scare me until after I'd left American shores. Since I was traveling overseas, my harassers knew they could batter me and I'd have no one to report to, particularly since I'd be embarrassed that I was being maligned through a dating site. Skeptics might naturally claim that I don't know for sure it was the Russians, and that's true—I don't have a video of the hackers and hecklers on the other end of a dating profile harassing me. But the account was overtly Russian, it bragged about being Russian, and the incident wasn't isolated: less than two months later, the old Gmail account attached to those dating apps showed attempted log-ins from places I had never been nor visited, and I received new waves of targeted phishing spam tailored specifically to me.

In August 2015, I dumped the Gmail account, cleaned my systems as best I could, and again operated under the assumption that everything I accessed on the internet was being watched either by the Russians, some terrorists, hackers, or all of the above. I started preparing for the worst: a Russian smear campaign fueled by *kompromat*, some true information about me mixed with a bit of false, all linked through discrediting narratives posted on blogs or news sites. I'd seen it happen to others around me, so, as I learned to do in the Army, I prepared for the worst and hoped for the best.

During the summer and fall of 2015, Weisburd, Berger, and I debated the merits of writing up an overview of Russia's use of English-speaking social media to influence Americans. But the Islamic State's violent growth and construction of a caliphate took center stage in national security circles. There seemed to be little interest in Russia and its online operations. Writing up our findings would likely draw significantly more cyberattacks against family and friends and result in few, if any, reads by the mainstream public and government officials. And then there was the issue of pay: we weren't receiving any support for this research, and all of us had let our curiosities hurt our pocketbooks at some point in the past. The costs seemed to far outweigh the benefits. But the troll army's behavior continued to evolve in the latter half of 2015.

Russia had decisively entered into the Syrian fight on Assad's behalf by the end of 2015. Looking for messaging strategies to derail the Islamic State and al-Qaeda's powerful radicalization narrative, I saw Russia as a possible target of jihadi angst, moving it away from the United States and Europe and toward a shared adversary. For two decades, al-Qaeda, and now its more virulent spawn the Islamic State, targeted the United States. These terrorists referred to America as the "far enemy" for its alleged support of apostate Muslim dictators, who were the "near enemy," in jihadi-speak. But the United States wasn't supporting the abusive assaults of the Assad regime and instead wanted the dictator out in Syria. Russia backed the Syrian regime, much in the way it had backed Afghanistan's dictatorship in the 1980s.

My thought was: why not remind jihadis fighting in Syria that it was Russia, and not America, backing the human-rights-violating President Assad? If successful in such messaging, the United States might potentially redirect jihadi narratives, much in the way I'd seen Russia's trolls point fingers at America.

On October 26, 2015, I posted on the Foreign Policy Research Institute's (FPRI) website a short blog entitled "Russia Returns As al Qaeda and the Islamic State's 'Far Enemy.'" It was the first time I'd written publicly about Russia's influence operations and, I figured, the last. Americans didn't appear to have much appetite or interest in anything other than the Islamic State, but this opinion piece offered an opportunity to discuss what had been a wasted effort to date.

The post received few reads and little attention, and I figured my remarks would drift into the morass of counterterrorism commentary filling social media feeds. But I was wrong.

Two weeks later, in November 2015, I checked my email messages as I walked into a friend's promotion ceremony outside Washington, D.C. FPRI's website administrator had sent a note regarding a visit he'd just received from an FBI agent.

I quickly dialed FPRI in Philadelphia. The FBI had detected a breach of the FPRI website, one that had installed sophisticated malware on my bio page and some of the articles I'd written. The FBI agent wouldn't say who was responsible for the hack, but the website administrator noted, "I think you were targeted because of what you write." *I bet so, and I bet I know who targeted me.*

Terrorists would badger me on social media, but few held the requisite hacking skills to breach computers halfway around the world, nor did they routinely have access to the more advanced malware found on the FPRI website. Hackers often target current and former government officials, but when it came to the Islamic State and al-Qaeda, collectives likely saw me as an ally against terrorists more than a foe. Cybercriminals could breach a system like FPRI's, but what did they stand to gain when they are motivated by money? A convoluted blackmail scheme against me using FPRI's website seemed quite far-fetched.

All signs pointed to a nation-state as the likely perpetrator. The FBI had detected the breach and malware installation fairly quickly by cybersecurity standards. They wouldn't reveal to me who had conducted the attack, but they seemed to be aware of the actor's techniques and had been scanning for the actor's breaches. The attack's timing, just weeks after my Russia article, made sense as well. Even more ominously, the malware that was installed provided hackers with a digital listing of all who accessed my profile and writings. This data would be useful for one purpose more than any others: the delivery of a discrediting campaign directly to the inboxes and social media feeds of those reading my opinions. The malware would provide the perfect mechanism for releasing alleged compromising information on me to those invested in my opinion. Such an attack on my credibility would weaken my public voice, discredit my claims, and shrink my influence among those who knew my work. The malware also provided a conduit by which a nation-state could target those most interested in resisting it. All evidence pointed to only one threat capable of, and motivated to pull off, such a cyber campaign—the same one powering the social media troll army I'd been watching. And again, I asked myself, *Is this really worth it?*

5

Harmony, Disharmony, and the Power of Secrets

"Brothers in Islam . . . they have influenced the International Community to believe that the Somali religious leaders are Al-Qaeda. . . . The following decision was made: . . . Cooperation has to be made with criminals and hard currency provided as motivations to assassinate the officials of the administrations . . . Care has to be maintained all along to avoid leaking of this information. Whosoever leaks this information and is found guilty should be shot. . . . Unity comes from Almighty Allah . . . Chief of the Imaam of the Islamic Courts . . . Shiikh Hassan Dahir Aweys (signed)."[1]

Sheikh Hassan Dahir Aweys, once the top leader of the Islamic Courts Union, provided the first sign of Shabaab's fractures when he defected in 2013.[2] Omar Hammami longed on Twitter for Aweys's support as he ran from al-Shabaab's assassins through the Somali forests. But Sheikh Aweys, years before his Shabaab escapades, surprisingly became the first victim of Julian Assange's creation, WikiLeaks.

WikiLeaks began its campaign for international transparency on December 28, 2006, posting a full English translation of a document allegedly written by Sheikh Aweys on November 9, 2005. The message made less news than the messenger. Julian Assange began his climb to international fame and today, more than a decade later, remains a disruptive force of information warfare. WikiLeaks thought "the crowd," an open-source army of contributors scattered around the world, would investigate the contents of the alleged Aweys letter and determine collectively whether the document was true to its alleged source. That never actually happened, though. Even today, the contents of the document are shrouded in mystery.

Assange's journey to WikiLeaks began where many transparency activists start: hacking. From his early years to the present, Assange has lived like a nomad, pursuing hacking as a hobby before turning it into his life's work. His teenage hacking led to criminal investigations and charges, a minor penalty, and later a fight with the Australian state for custody of his child. Whether it was child custody battles with Australia's Department of Health and Community Services or criminal prosecutions, Assange developed a complete disdain for institutional hierarchies and patronage networks. He crafted his own manifesto, entitled "Conspiracy as Governance," claiming that illegitimate governance is conspiratorial and the product of people working in "collaborative secrecy, working to the detriment of a population." This manifesto guided Assange's vision of WikiLeaks, a virtual hub of secret documents, leaks to be used in information attacks against the corrupt and the criminal, the states, organizations, and bodies they govern.[3]

Raffi Khatchadourian's 2010 *New Yorker* article "No Secrets" provides an exceptional accounting of Assange and the early years of WikiLeaks. In 2006, Assange invited collaborators to the WikiLeaks mission: "Our primary targets are those highly oppressive regimes in China, Russia and Central Eurasia, but we also expect to be of assistance to those in the West who wish to reveal illegal or immoral behavior in their own governments and corporations."[4] Assange remarked passively that his social movement to expose secrets could "bring down many administrations that rely on concealing reality— including the US administration." Rop Gonggrijp, a Dutch activist, hacker, businessman, WikiLeaks funder, and overall backbone for Assange's operations, told Khatchadourian that WikiLeaks plays an essential role in the media. "We are not the press," Gonggrijp said. Rather, he considered WikiLeaks to be an advocacy group for sources. According to Gonggrijp, WikiLeaks created a world in which "the source is no longer dependent on finding a journalist who may or may not do something good with his document."

The statements by both Assange and Gonggrijp appeared almost immediately to be at odds with each other and with reality. WikiLeaks wasn't the press, but it would provide raw information to all of the press in hopes that someone would get the story "right" by WikiLeaks' standards. WikiLeaks would go after the most oppressive regimes and any behavior it deemed illegal or immoral, making itself the arbiter for the world as to where the blurry lines of morality lie. These paradoxical statements have played out in confusing ways throughout WikiLeaks' history.

Notoriety brought in new pilfered secrets, and Assange claimed in 2010 that WikiLeaks was receiving dozens of disclosures a day. Each year, these disclosures struck bigger targets with larger caches of pilfered materials. WikiLeaks' hit list from 2006 to 2009 included China, Kenya's police force, Scientology, Sarah Palin, Bank Julius Baer, the Bilderberg Group, and Iran. But gradually, Western governments and businesses began to outnumber oppressive regimes by a sizable margin, and Assange's top target increasingly became the U.S. government.

In early 2010, an Army private first class then named Bradley Manning provided WikiLeaks with hundreds of thousands of diplomatic cables and classified reports, along with a 2007 video showing a highly contentious Baghdad airstrike by American Apache helicopters that had killed twelve people, including two Iraqi journalists working for Reuters.

WikiLeaks and Assange released their "Collateral Murder" video, using footage delivered by Manning, creating a worldwide debate about not only the contents and context of the video but the need for justice and accountability and the legality of the attack. The video appeared to be a major step forward for WikiLeaks' pursuit of injustice. Donations poured in, awards came from human rights groups, and Manning, the source of the videos, went to military prison, with a thirty-five-year sentence.

Assange's journey since 2010 has been the substance of documentaries, movies, articles, and books. Internet hosting providers waxed and waned as Western governments, particularly the United States, placed enormous pressure on WikiLeaks' technical backbone, seeking to take the outlet offline. Until the U.S. presidential election of 2016, surprisingly few questioned the validity of Assange's attacks on the West and particularly on the United States.

Most overlooked a curious bit of WikiLeaks history, the first glimpse of which occurred on November 17, 2009. WikiLeaks posted email messages between climate scientists at the University of East Anglia's Climatic Research Unit (CRU). The emails in the raw were used by climate change skeptics to show global warming to be a conspiracy. The CRU claimed that the emails were nothing more than healthy dialogue between researchers. Some investigating the CRU's breach thought the leaks may have come from Russia, noting signatures that could have been tracked back to "a small web server in the formerly closed city of Tomsk in Siberia."[5] The source of the hacks remains an unsolved puzzle but the suggestion of a connection between Russia and WikiLeaks, curiously, would surface again less than a year later.

"We have [compromising materials] about Russia, about your government and businessmen. . . . We will publish these materials soon," Assange said during an interview with the pro-Russian-government daily newspaper *Izvestia*, in what appeared to be a dire warning to Moscow. Kristinn Hrafnsson, another WikiLeaks spokesperson, repeated the warning on October 26, 2010: "Russians are going to find out a lot of interesting facts about their country."[6] Audiences and journalists waited in anticipation for the Russia bombshells, but they never came.

The following day, October 27, 2010, an unnamed official at the FSB's Center for Information Security, Russia's internal intelligence arm, issued a statement: "It's essential to remember that given the will and the relevant orders, [WikiLeaks] can be made inaccessible forever."[7]

The Russian secrets never surfaced at WikiLeaks, and instead Assange's next posting, on November 28, 2010, showcased U.S. State Department stolen diplomatic cables, beginning the slow drip of roughly 250,000 reports harming U.S. relations with countries worldwide. WikiLeaks' challenge to the worst regimes, and Assange's bravery in the face of dictators, faded away.

Israel Shamir, a close associate of Assange's, also began appearing in WikiLeaks circles in 2010. James Ball, a staffer at WikiLeaks during a tumultuous three-month period, described Shamir's entrance in WikiLeaks circles: "A self-styled Russian 'peace campaigner' with a long history of anti-Semitic writing . . . Shamir was introduced to the team under the pseudonym Adam, and it was only several weeks after he had left—with a huge cache of unredacted cables—that most of us started to find out who he was."[8] A little while later, Shamir landed in Belarus, a Russian ally led by Alexander Lukashenko, who has held on to power through press censorship, communications monitoring, and, above all, the manipulation of politics to quell dissent.

Shamir allegedly provided a cache of unredacted American diplomatic cables to Lukashenko's chief of staff, Vladimir Makei. Kapil Komireddi, of the *New Statesman*, said that Shamir then "stayed in the country [Belarus] to 'observe' the presidential elections."[9] Lukashenko won the vote on December 19, 2010, by an overwhelming majority, one so suspiciously lopsided that mass protests by Belarusians led to the dispatching of the state militia to restore calm. In January 2011, the Belarusian state-sponsored newspaper *Soviet Belarus* published extracts of U.S. diplomatic cables provided by Shamir and WikiLeaks. Among those exposed as recipients of foreign cash were a defeated opposition candidate to Alexander Lukashenko, Andrei Sannikov. Also outed in the published cables were Sannikov's press secretary, who'd died suspiciously months before, and another Lukashenko political opponent, Vladimir Neklyayev, was subsequently placed under house arrest.

Criticisms of Assange's connections to Shamir mounted in late 2011. WikiLeaks supporters became rightly confused as to why the transparency outlet had helped Lukashenko, the very type of authoritarian the group had originally sought to target. Then, on April 17, 2012, Julian Assange's show, *World Tomorrow,* debuted on Russia's state-sponsored news outlet RT broadcasting twelve episodes into the summer of 2012. Two months after that, Assange retreated to the Ecuadorian embassy in London, to avoid extradition on Swedish rape charges. The Swedish charges have been dropped, but Assange remains inside the Ecuadorian compound, where, ever since October 27, 2010, his efforts have supported the agenda of one country above all others: Russia. And WikiLeaks' document dumping has harmed one nation most of all: the United States.

\\\\\\\\\\\\\\\\\\\\\\\\\\\\\

Al-Qaeda first felt the American response to the September 11 terrorist attacks on October 7, 2001.[10] Each guided missile that struck an al-Qaeda or Taliban target drove veteran mujahideen from their safe houses and scampering into the mountains. U.S. Special Forces soon arrived in droves. They teamed with the Northern Alliance and sprinted across the country in pursuit of Osama bin Laden and his top deputies. Some terrorists were killed right away, and many others left so quickly that secrets tucked in al-Qaeda's safe houses and stored on hard drives fell into the hands of advancing American commandos and intelligence officers.

Gregory Johnsen, a PhD scholar of the Middle East, accurately noted in his studies of al-Qaeda in the Arabian Peninsula that bin

Laden was like many other Arab volunteers to the Afghan mujahideen but for one exception: he had money.[11] The son of Mohammed bin Awad bin Laden, founder of the Saudi Binladin Group and the wealthiest non-royal Saudi at the time of his death, Osama came from wealth no other jihadi could match.

Osama bin Laden briefly followed in the footsteps of his father, operating a construction company before traveling to Pakistan to join the mujahideen resistance to the Soviet Union. Bin Laden used his business acumen to form the Afghan Services Bureau (Maktab al-Khidamat) in Peshawar, Pakistan. His logistical skills proved more valuable than his time in combat. While he briefly fought in the Battle of Jaji, winning praise from the more hardened battlefield soldiers, his tracking of personnel and expenditures became the backbone on which he built a global front—the name al-Qaeda is Arabic for "the Base"—that would take jihadi activity to an entirely different level. Bin Laden's vision, inspiration, money, logistical support, and organizational coordination brought him and top veterans from Afghanistan to command a global legion fighting in nearly every Islamic war in the world during the 1990s. When the Twin Towers fell on September 11, 2001, bin Laden commanded a global terror network operating on at least four continents and was responsible for the most devastating attack on U.S. soil since the Japanese raid on Pearl Harbor, nearly sixty years before.

Bin Laden's mastery of planning and logistics offered a significant advantage over competing jihadi leaders and their nascent extremist groups during the 1990s. He and his deputies went to great pains to document and archive the actions of his group. Al-Qaeda tracked every expense and logged correspondence. This meticulous tracking served several purposes: it allowed bin Laden to efficiently run his organization, sustain operational control over a disparate network, and create a running historical archive for how his team sought to change the world. It's unlikely that bin Laden ever appreciated the downside of keeping detailed notes on his activities.

It might be surprising to many that before the United States got wikileaked by WikiLeaks, the United States wikileaked on al-Qaeda a few months earlier, in February 2006. Bin Laden's ledgers and communiqués lay in tatters or etched on hard drives confiscated during raids and battle damage assessments. Each successful strike provided not only the keys to where another al-Qaeda operative resided but a diagram to al-Qaeda's internal workings, its strengths and vulnerabilities. The evidence traced how bin Laden's team had bridged the transition from analog to digital. Charred handwritten notes dating back to the group's founding, in the early 1990s, littered the scenes of targeted bombings. Computers, CD-ROMs, and floppy disks sat in safe houses, holding terrorist personnel records and employment contracts like those you'd find in any American corporate human resources department.

In the early days of the war on terror, U.S. counterterrorism focused on the immediate and the tactical. The pursuit of bin Laden, killing or capturing his deputies, and preventing the launch of another September 11–style attack were the obsessions of America's wide-ranging manhunts. As American forces pushed al-Qaeda's remnants in Afghanistan into Pakistan, Iran, and a smattering of Middle Eastern locales, the sheer volume of records recovered by U.S. forces stretched from the thousands to the hundreds of thousands and then millions of documents. Storing and archiving al-Qaeda's secrets as thousands of American troops scoured the earth in pursuit of terrorists became a challenge—one requiring some method for centrally storing and recording what had been unearthed.

In 2002, CIA director George Tenet created the Harmony database, as the intelligence community's central repository for all information gathered during counterterrorism operations. The database served as a single place for bringing together information for the conduct of the emerging field of intelligence known as DOMEX—document and media exploitation. Similar to the way al-Qaeda centralized documents for a sprawling terror network, the Harmony database centralized documents being recovered by a sprawling counterterror network. At first, the Harmony database assisted soldiers picking up clues about enemy whereabouts and communications from many different battlefields and helped support the prosecution of alleged terrorists.[12]

As documents populated the Harmony database, bin Laden's trail went cold in Pakistan. His videotaped messages surfaced on Al Jazeera, a new franchise burst onto the scene in Iraq, and new internet forums were uniting a young generation of terrorist recruits. By 2005, the rapid fall of Afghanistan and the collapse of Saddam's regime in Iraq had brought not the end of al-Qaeda but a new chapter. Rapid, rampant military pursuit of terrorists everywhere hadn't rendered the group defunct. Instead, al-Qaeda surged back to life, invigorated by the violent rampages of Zarqawi and connected via chatrooms and internet forums. Raids on al-Qaeda operatives in Afghanistan and Iraq yielded the usual intelligence on the "next" target but didn't provide much perspective on how to defeat the decentralized terrorist menace responsible for the September 11 attacks.

While most of the military focused narrowly on the raging battles in Central and Southwest Asia, U.S. Special Operations Command (SOCOM), headquartered at MacDill Air Force Base, in Tampa, Florida, had received the broader mission of pursuing al-Qaeda, its affiliates, and its associates anywhere they might be, using any means available. Military planners at combatant commands frame their comprehensive strategic approaches to winning battles with the acronym DIME. Diplomacy, information, military, and economics represent the four strategic components for structuring well-rounded campaigns. Bombs, bullets, raids, and renditions didn't muffle jihad's call to kill Americans; more military muscle truly couldn't be applied. Cutting off terrorist funding sources hadn't made much of an impact, either, and economically al-Qaeda operated on a shoestring budget in most theaters. Diplomatic efforts had brought many partners into the U.S. coalition to counter terrorists, but al-Qaeda, as a non-state actor, didn't have any counterpart for negotiation, and if even if it had, negotiation wasn't an option for America.

The one aspect of national power where the U.S. military trailed al-Qaeda was information. Amid the United States' renewed push toward public diplomacy at the national level, a smart officer at SOCOM understood the information war better than others. He gazed upon the Harmony database and saw in the millions of documents something different from those looking for tactical bread crumbs to track al-Qaeda's movements. Instead, Major Steve saw al-Qaeda's secrets from a different perspective. Strewn throughout their files and hard drives were the strategic deliberations of terrorists, their biases and preferences, expense reports, likes and dislikes, and successes and failures, as well as what they thought of one another. Al-Qaeda's documents in sum yielded insights into the group's strategic weaknesses and internal fractures. The military's intelligence apparatus was great at tagging, tracking, and demolishing targets, but it was never meant to divine the historical, religious, and cultural trends, the strengths and weaknesses of al-Qaeda. Major Steve had an idea who could figure this out and, oddly, it was the same solution proposed by WikiLeaks: the "crowd."

The Combating Terrorism Center at West Point, where I worked, offered an interface for the military and government to connect with top experts in the new cultures, regions, languages, and politics challenging effective counterterrorism operations. Major Steve recognized the opportunity the CTC could offer. At West Point, he could unlock the Harmony database's secrets, create an open-source repository for the public, and enlist highly educated military officers stationed at West Point to study and collaborate with top professors positioned around the world. In 2005, the CTC launched the Harmony Program "to contextualize the inner-functioning of al-Qa'ida, its associated movement, and other security threats through primary source documents." In addition to conducting initial research on the materials, the program aimed "to make these sources, which are captured in the course of operations in Iraq, Afghanistan and other theaters, available to other scholars for further study."[13]

The center's faculty analyzed the first study and release of supporting documents, entitled *Harmony and Disharmony: Exploiting al-Qa'ida's Organizational Vulnerabilities*, on Valentine's Day 2006. Cable television specials and endless news shows since September 11, 2001, had pushed the belief among Americans that al-Qaeda's operations were menacingly flawless and that the organization operated at near perfection. But this first release of the group's internal documents revealed it to be somewhat ordinary and remarkably similar in its administrative functions. Al-Qaeda's employment contract appeared to be no different from that of an ordinary American worker. Theological debates and struggles with how to secretly control terrorist operatives shone through in most every document. These conversations offered opportunities for the United States to exploit al-Qaeda's weaknesses.

The employment contract showed that Arab recruits were paid more than African recruits, and married volunteers with children received more benefits and vacation than single members. The report noted that ineffective terrorists, rather than be plucked off the battlefield, "should not be removed from the network if they can be reliably observed, even if they present easy targets."[14] The report's justifications for this recommendation pulled from a 1999 email sent by Ayman al-Zawahiri to a Yemeni cell leader in which he scolded a subordinate, saying, "With all due respect, this is not an accounting. It's a summary accounting. For example, you didn't write any date, and many of the items are vague."[15] Nearly twenty years later, Zawahiri's letter offers some insights into why terrorists in the ranks sought to defect to ISIS after bin Laden's death: he was a stickler of a boss.

The release of the Harmony documents directly supported the report's key recommendation: "increase internal dissension within the al-Qa'ida leadership."[16] Strewn throughout the documents was the organization's dirty laundry. Communiqués between al-Qaeda subordinates challenged the direction put out by the group's leaders and questioned whether orders should be obeyed. One member said that faulty leadership held the group back, asserting that bin Laden had rushed "to move without vision," and asked Khalid Sheikh Mohammed, mastermind of the 9/11 attacks, to reject bin Laden's orders.[17]

The CTC's findings were some of the first to analyze al-Qaeda holistically and offer some alternatives to the previous five years of war. The insights published by West Point's scholars were good, but more important were the audiences engaged by the released documents: the public "crowd" and the terrorists in al-Qaeda's ranks. At the time of the *Harmony and Disharmony* report, there were more people studying terrorists than there were terrorists, and very little of this research pulled from internal documents or primary sources.

The release of the Harmony database into the wild had suddenly sparked new research efforts around the world. The original Arabic documents were included with English translations. A database at the Combating Terrorism Center's website hosted the files, and hits on the servers showed how quickly word spread. Scholars with regional expertise, language skills, and in-depth knowledge of Arab cultures scoured the documents and returned deeper insights. Journalists used bin Laden's and his team's communiqués as the basis for investigating news stories. Everyone around the world seeking to end the terrorist group that had perpetrated the horrors of September 11 was suddenly more empowered. The "crowd" could help bring about the demise of al-Qaeda.

Al-Qaeda terrorists came to look at the Harmony documents as well. Those who'd survived America's endless barrage now worried whether their private messages trash-talking top terror leaders would result in their purge from within. Among the Google Analytics for the Harmony Project were a growing list of pings from anonymous web users, people employing proxy servers to mask their online identity. The time zones for these pings suggested that they resided on the other side of the world. Al-Qaeda's secrets were now working against them.

Harmony and Disharmony's success led to two more Harmony database reports the next year. *Al Qa'ida's (Mis)Adventures in the Horn of Africa's* first report released a second tranche of documents focused on al-Qaeda's early operations in East Africa, between 1992 and 1994, when the terror group tried and failed to infiltrate Somali clans. The remnants of this expedition later seeded the operational cells in the Horn of Africa that conducted the 1998 U.S. embassy bombings in Kenya and Tanzania.[18] The second report's primary findings were that al-Qaeda fared better not in failed states like Somalia, as conventional wisdom might suggest, but in weak states like Kenya and Pakistan, where weak national security operations restricted American counterterrorism efforts. Next came *Cracks in the Foundation*, which studied three periods of al-Qaeda's history using the group's internal documents.[19] The study found that al-Qaeda, as a military organization, had never been particularly strong, and its success as a media organization masked deep internal divides between its leaders over strategic direction.[20]

The Harmony database armed crowds of researchers scattered around the world with more secrets for studying how to divide and conquer al-Qaeda's terrorists. The Harmony documents surfaced in theses, dissertations, books, and government publications and even among terrorists. And then there was poor Sheikh Aweys. Only a few months after WikiLeaks' December 2006 launch with the release of the Sheikh Aweys assassination order, an application form for a Kenyan visa surfaced among the al-Qaeda files released in the second Harmony report. The application read, "Hassan Dahir Haji Aweis . . . Date of Birth: 1944 . . . Country of Residence: Somalia . . . Reasons for Entry: Business."[21] The first guy outed by WikiLeaks, the veteran jihadi with whom Omar Hammami longed to connect on Twitter—his visa application from September 3, 1993, surfaced in a pile of al-Qaeda documents recovered in Afghanistan in 2002 showing his connections to the world's most prolific terrorist group. Within a year, whether it was WikiLeaks or the Combating Terrorism Center, Sheikh Aweys had been outed by two transparency efforts that used two different methods to bring his secrets into the public.

\\\\\\\\\\\\\\\\\\\\\\\\\\\\\\\

"Can you tell me a secret, or would you then have to kill me?" Americans love plays on this iconic phrase, born in books, that has migrated to endless television shows and movies about spies and espionage. Like, I imagine, most every government person who has ever held a security clearance, I've received this silly query thousands of times. My response, before I force a laugh to push away the question, is always the same one, which I never utter: "Yes, but if I told you a secret, would you even understand it?" Calls for transparency have risen proportionally as more information has moved to the internet. More secrets have spilled into the open in the past decade than anytime before. But do we understand more, or less, after all these disclosures?

Democracies and the citizens within them have always pursued the virtue of openness. But political scientists James R. Hollyer, B. Peter Rosendorff, and James Raymond Vreeland rightly point out in their study "Democracy and Transparency" that empirical analyses of the topic show "the policymaking of democratic governments is shaped by transparency and, importantly, democratic governments have incentives to obfuscate evidence."[22] Citizens want transparency, and elected officials want to keep their jobs.

WikiLeaks and the U.S. government both launched transparency initiatives in 2006, and while they mirror each other in concept, they're distinctly different from each other in execution. In fact, they are diametrically opposed in terms of methods and declared intentions.

WikiLeaks doesn't hack to gain secrets, but relies on others to provide them with secrets they can disclose. With the declared intention of fighting corruption and authoritarians, WikiLeaks matches and pulls from hackers and insiders whose personal grievances align with the group's stated grievances. This is where WikiLeaks' methods are oddly misaligned with the declared intentions of those who provide them with secrets. The couriers of WikiLeaks secrets, at least for their big public disclosures, arise not from the most corrupt, oppressive regimes in the world, but the most open, for the consequences of these data thefts in the former is death, and in the latter fame.

Two Americans, Chelsea Manning (the former Bradley Manning) and Edward Snowden, remain the most famous couriers connected to WikiLeaks. The former fed the outlet; the latter was assisted by it. Their disclosures surfaced in different places, but in spirit they sought the same goal: the transparency and accountability of the U.S. government. Both insiders were young at the time of their insider breaches, in their twenties. Both spent only a short time in their government roles before spilling the beans on their employer, the U.S. government.

Manning began basic training on October 2, 2007, before arriving in Fort Drum, New York. Manning had a difficult time in training, and she found herself sitting in Baghdad by October 2009. In January 2010, only three months after arriving in Iraq, she contacted WikiLeaks.

WikiLeaks acted not as an outlet for Edward Snowden, but a shepherd. After Snowden departed for Hong Kong in May 2013, having stolen vast collections of highly sensitive U.S. intelligence, his next move was not to stand trial and be exonerated by the public for his breach, but to seek safe harbor. Snowden needed help, and WikiLeaks stepped in, dispatching Sarah Harrison, of WikiLeaks' legal defense team, to Hong Kong, where she then traveled alongside Snowden as he made his way to Moscow.

Snowden's story prior to his breaches moves through professional peaks and valleys. An enlistment in the Special Forces in 2004 ended with injury and discharge later that year, due to broken legs, according to Snowden. Recruited into the CIA as an information technology specialist in 2006 and stationed in Geneva in 2007, Snowden, after some professional success, decided to leave the CIA in 2009. He claims his thinking and faith in the CIA changed while in Geneva. Or was it something else? The *New York Times* reported on October 10, 2013, that, "just as Edward J. Snowden was preparing to leave Geneva and a job as a C.I.A. technician in 2009, his supervisor wrote a derogatory report in his personnel file, noting a distinct change in the young man's behavior and work habits, as well as a troubling suspicion."[23] Years later, a congressional report released findings of an investigation of Snowden's hacking:

> Snowden was, and remains, a serial exaggerator and fabricator. . . . He claimed to have left Army basic training because of broken legs when in fact he washed out because of shin splints. He claimed to have obtained a high school degree equivalent when in fact he never did. . . . He also doctored his performance evaluations and obtained new positions at NSA by exaggerating his résumé and stealing the answers to an employment test.[24]

Did Snowden steal secrets in support of an ideology of transparency, to make Americans aware of government-authorized surveillance and the need for more accountability? Or was it revenge for disciplinary action, or an obsessive, ego-driven disclosure in pursuit of fame and acknowledgment? Or was it "all of the above"? If Snowden was so convinced that his actions were merited, why not stand trial in the United States and seek to be exonerated by those Americans he sought to protect?

Others before Snowden have disclosed secret U.S. government documents in pursuit of the public good. The *New York Times* in 1971 published the Pentagon Papers, detailing secrets about the conduct of the Vietnam War, and exposed glaring discrepancies between the White House's declared scale of the Vietnam conflict and the reality in Southeast Asia. Daniel Ellsberg, a RAND researcher who had worked on the study, served as the courier for the secrets. The U.S. government charged Ellsberg with conspiracy, espionage, and theft of government property, but these charges were later dismissed and he was exonerated.

The confusing intentions of WikiLeaks' leakers blend in with the contradictions of WikiLeaks' methods. In 2010, Assange bragged about the rapid rate of disclosures landing in WikiLeaks inboxes, and yet not even a fraction of these leaks have appeared on its website. Assange and his team decide what's published and what's not, and in so doing set an agenda and show their intentions. Not only does WikiLeaks decide what's released, but when and how much of each document dump. Assange's complaints about journalists not using disclosures effectively land hypocritically when WikiLeaks releases stolen cables and emails to prioritize one narrative against another.

Assange, as the controller of undisclosed secrets, has built an elitist information empire, can and does manipulate perceptions, and is now the governor of "collaborative secrecy, working to the detriment of a population," as he wrote in his own manifesto.[25] Assange is his own worst nightmare—he is "conspiracy as governance." He's become another version of what he claimed to have set out to destroy.

Furthermore, WikiLeaks has a growing tendency to be both the source of leaks and their narrator. Assange voiced frustration when nothing came of the "Collateral Murder" video. "This was such a fucking fantastic leak: the Army's force structure of Afghanistan and Iraq, down to the last chair, and nothing," he said.[26] The *New Yorker*'s Khatchadourian wrote that Assange told him "his mission is to expose injustice, not to provide an even-handed record of events."[27]

A staple of WikiLeaks since its inception has been trying to convince the world of Assange's philosophy and interpretation of facts, but that vision has become increasingly difficult to comprehend. Assange's frustration with people not understanding his point of view has led him to add further commentary to WikiLeaks disclosures. Over the years, Assange has become an expert in everything he discloses—climate change, military operations, and the political processes and cultures of countries stretching from Africa to America. His celebrity has grown as he doles out his opinion; his influence has shaped debates.

This dynamic has been further exacerbated by social media. Holed up in the Ecuadorian embassy in London, Assange advances narratives surrounding released information, but extraordinarily few who read his tweets examine the actual underlying leaked documents. Assange provides his context to the disclosures, which Twitter users routinely accept as truths when they are actually opinions. Furthermore, WikiLeaks and Assange don't seek to avoid harming innocent bystanders in their disclosures. Assange called it "collateral damage, if you will," claiming he doesn't have time to weigh the importance of every detail in every document. This seems strange, given that his team seems to have plenty of time to determine which documents to publish and which to keep out of the public eye.

Unlike WikiLeaks, no one really knows who the exact couriers of bin Laden's secret archives were. But it's no secret how the Harmony documents arrived in the database: airstrikes in Afghanistan and the raid on the Abbottabad compound were televised for all to see. The couriers may be unknown, but the method is not. Disclosures from West Point's Combating Terrorism Center, like those from WikiLeaks, remain selective, centered around topics rather than massive caches of documents on a wide range of topics.

Releases from the Harmony database were slow, and deliberately so, for a couple of reasons. The Department of Defense sought to ensure that the personal data of those innocently captured in the records didn't cause undue harm. Most of all, much-needed context accompanied the release of the documents. Experts in the disciplines, topics, cultures, and countries covered in the records provided the public with a baseline of clarity for what was made available, what it might mean, flaws in the data and its collection. The narrative sought not to restrict further research but to provide a context that prevented harm to others and sought to quell poisonous false conspiracies pulled from kernels of truth merged with partisan conjecture.

Above all, no one doubted or questioned the intentions of the Harmony effort. They were clear. The intention was to stop violence, to help others stop violence, not to pick a side or advance a narrative. The more that was disclosed, the more everyone benefited, the U.S. government included. If other researchers arrived at conclusions different from those of the Combating Terrorism Center, even better.

Meanwhile, WikiLeaks inspired others oppressed in truly corrupt and criminal regimes to follow their methods. As terrorists moved to blogs and social media, those terrorized by oppressors tried their hand at transparency, to shine light where darkness had long prevailed. Some failed, and others succeeded, and while WikiLeaks went in a strange direction, others with similar intentions as those stated by Assange followed methods closer to those of the Harmony Project and achieved results neither WikiLeaks nor the Harmony Project could have anticipated.

\\\\\\\\\\\\\\\\\\\\\\\\\\\\\\\\

Along Mexico's northeastern border with the United States, residents remain trapped between a corrupt government and criminal drug cartels. Newspapers, in free societies, act as a traditional check on corrupt governments and provide an accounting of violent behavior. But life in Mexico is anything but free, and newspapers with bylines and sources have frequently felt intimidation from drug cartels and the government. Those journalists pushing the envelope of accountability and challenging the strong rarely last long. Nearly one hundred Mexican journalists have been killed in the past two decades covering the life and times of Mexicans. Again, social media lowered the barrier to entry for the common man, and soon an alternative to traditional journalism emerged to account for the state of violence in Mexico. People lacking a way to defend themselves turned to blogs, forums, Facebook, and Twitter to reveal the secret lives, the real world, of Mexico's drug cartels.

Two blogs, *Al Rojo Vivo* and *Blog del Narco*, broadcast reports of violent incidents. Another site, *Nuevo Laredo en Vivo*, detailed the criminality and oppression, exposing what citizens in the region encounter every day trying to survive. Social media accounts on Twitter and Facebook shared updates from these blogs. The posts offered up transparency with regard to life in Mexico but also provided catharsis for those posting anonymously. Each revelation served as a small way for those without power to fight back against a world and a way of life they didn't deserve.

Awareness of the violent region grew throughout 2011, and comments and views increased on blogs documenting the violence. Those posting anonymously online felt secure, for a time. But the blogs' popularity quickly drew the attention of those inclined to suppress them.

In Nuevo Laredo, the drug cartels don't fear the police—the police fear the drug cartels. While the American media ran endless coverage on the tenth anniversary of the September 11, 2001, terror attacks, citizens of Nuevo Laredo woke to a common grisly sight, with a new twist for the social media era.

Hanging from a pedestrian bridge was a hogtied, disemboweled woman. Her nude torso, with protruding intestines, was displayed alongside a man dangling by his hands, with his arm nearly torn from his body. The fingers and ears of the man and woman were badly mutilated. "This is going to happen to all of those posting funny things on the Internet," read a sign posted nearby. It was annotated with the letter *Z*, suggesting the murders were the handiwork of the locally dominant Zetas cartel.[28]

Bloggers, commenters, and social media personas took to forums and called on the online resistance to rally. "Don't be afraid to denounce. It's very difficult for them to find out who denounced. They only want to scare society," said an anonymous person on the *Al Rojo Vivo* forum."[29]

But it really wasn't that difficult for the cartels to determine who the sources of dissent were in Nuevo Laredo. Less than two months later, another body turned up. "Hi, I'm Rascatripas and this happened to me because I didn't understand I shouldn't post things on social networks." The message appeared on a blanket under a blood-soaked, decapitated body. The body was the fourth in Nuevo Laredo in three months, and bloggers on *Nuevo Laredo en Vivo* claimed that the decapitated man wasn't part of their forums, that the death represented nothing more than a scare tactic. But in a world of anonymous sources and cartel intimidation, how would either side even know? The Mexican social media activists continue on, but for them, the consequences remain great, and the outcome is hard to measure. They've united online, but not on the streets, and the world has not come to help them.

\\\\\\\\\\\\\\\\\\\\\\\\\\\\\\\\\

The Panama Canal bridges the land gap between the Atlantic and Pacific Oceans, and the law firm Mossack Fonseca, its office in Panama, provides a bridge between open markets and dark money. The firm, started by the German Jurgen Mossack in 1977, had risen to become the world's fourth-largest offshore law firm. By 2016 it was operating a global network of forty-two franchises, with significant subsidiaries scattered in the world's most advantageous tax havens. The firm catered to the rich, the powerful, and those seeking to keep their transactions private. That is, until April 2016.

The largest leak of confidential information ever hit the presses on April 3, 2016, when 2.6 terabytes of data, containing 11.5 million documents from Mossack Fonseca, were obtained by the German newspaper *Süddeutsche Zeitung* and then shared with the International Consortium of Investigative Journalists (ICIJ). The leak, referred to internationally as the Panama Papers, dwarfed any previous disclosure by WikiLeaks. The data dump differed from WikiLeaks not only in size but in approach. Rather than post the raw data without context or with Julian Assange's narrative, ICIJ shared the data with dozens of news outlets around the world familiar with the characters appearing in the documents and how the contents made sense in each country. In just the first eight months after their release, the Panama Papers generated more than 4,000 stories from media outlets and more than 6,500 investigations into companies and people potentially seeking to skirt laws and avoid taxes.[30] Not only did the Panama Papers achieve far more transparency than WikiLeaks, but they exposed an authoritarian regime that WikiLeaks initially set out to challenge but mysteriously never touched: Russia.

The Panama Papers revealed the dark money dealings of 143 politicians and twelve national leaders. Twenty-three individuals sanctioned for supporting regimes in North Korea, Zimbabwe, Russia, Iran, and Syria surfaced in the records. Above all, the *Guardian* noted, "a $2bn trail leads all the way to Vladimir Putin. The Russian president's best friend—a cellist called Sergei Roldugin—is at the centre of a scheme in which money from Russian state banks is hidden offshore."[31] Deeper analysis of the records suggests that the law firm represented one step in a shell game of financial maneuvers and false deals by which money moved out of Russia and then back into the hands of Putin's close friends and family—money and transactions that never bore Putin's name.[32] The Panama Papers shared one thing with the disclosures of Edward Snowden: the organization whose information was stolen wasn't breaking the law. Mossack Fonseca provided fully legal services.

The ICIJ followed up in 2017 with another bombshell, the Paradise Papers. This time, 13.4 million documents focused on the Bermudan law firm Appleby. Again ICIJ received documents and then siphoned them to "more than 380 journalists from over 90 media organizations in 67 countries," who spent months analyzing the documents before release.[33] The Paradise Papers showed how Appleby had helped "clients reduce their tax burden; obscure their ownership of assets like companies, private aircraft, real estate and yachts; and set up huge offshore trusts that in some cases hold billions of dollars." Whereas the Panama Papers implicated global players, the Paradise Papers struck closer to the United States. ICIJ repeated its formula, revealing significant Kremlin funding to an investor with stakes in both Twitter and a real estate technology company founded by President Donald Trump's son-in-law, Jared Kushner. Wilbur Ross, President Trump's commerce secretary, was shown to have investments in a company called Navigator Holdings, which had ties to Russian oligarchs.

The Panama Papers and the Paradise Papers revealed the power of leaks, but they also demonstrated the importance of context and analysis when communicating findings to the public. The disclosures were also different from WikiLeaks' because they targeted regimes that would fight back. Assange, who likes to think of himself and his colleagues as vigilantes for justice, struck only the easiest targets: Western democracies. Assange and his associates claim persecution and victimhood, but none of them have suffered the fate of some of those who covered the Panama Papers.

Politico described Daphne Caruana Galizia as a "one-woman WikiLeaks, crusading against untransparency and corruption in Malta, an island famous for both."[34] Galizia led the Panama Papers investigation in Malta, exposing corruption tied to the country's government. In 2017, she linked Maltese prime minister Joseph Muscat and his aides to offshore companies and payments from the government of Azerbaijan.[35] Her blog gathered up to "400,000 readers, more than the combined circulation of the country's newspapers" on a good day.[36] She chased everyone reeking of corruption: the government, the gaming industry, and the Mafia. That is, until Monday, October 16, 2017. Less than an hour after publishing a blog post, Galizia arrived at her car, a Peugeot, and a short time later, a powerful explosion tore through the vehicle, killing her instantly.

More than a decade after WikiLeaks' inception, the outcomes and consequences of its efforts seem unclear. Those originally designated as targets have benefited the most and, rather than transparency, the organization has sown confusion.

The Harmony Project wasn't perfect, but it was an example of how to provide context with content, how to make secrets digestible for the public and not fodder for fake news. Ultimately, the Paradise Papers and the Panama Papers created far more transparency and accountability than any of WikiLeaks' data dumps. One can argue about the ethics of disclosing a breach of personal information obtained by someone breaking the law, but the method by which the ICIJ conducted its disclosures through knowledgeable filters informing the public rather than confusing it has brought more accurate transparency to the public. The Harmony Program did the same, empowering citizens of every nation to disrupt the violence of jihadis and striking fear in the mind of bin Laden.

WikiLeaks' increasingly partisan information bombs designed to harm Western democracies lack context and instead play into the hands of manipulators, those increasingly empowered by social media. WikiLeaks' caches regarding climate change and the Bilderberg Group have created an outlandish string of false information and social media conspiracies. To be fair, WikiLeaks has won some awards and provided some perspective, and, by extension, its assistance to Snowden has raised awareness of government surveillance—the National Security Agency years later abandoned one of the programs Snowden had disclosed. But have the gains outweighed the costs? Snowden's leaks and the exaggerated claims of his accomplices, who pushed a narrative as false as it was true, have created high levels of distrust among Americans for their government, eroded confidence in elected officials, and damaged Western alliances. These transparency initiatives haven't strengthened democracy but tarnished it. They've not helped Americans or the West understand complex issues, but they have helped authoritarians rise as free societies decline.

October 27, 2010, proved a tipping point for transparency, when Assange went from a pioneer to a puppet. Others have traveled in his footsteps, couriers have supported him, and for them the consequences have been far higher. Chelsea Manning went to prison, and Assange called for her release, going so far as to say he'd accept extradition to the United States if Manning were granted clemency. President Obama commuted Manning's thirty-five-year sentence before leaving office in January 2017. More than a year and a half later, Assange remains in the Ecuadorian embassy, not honoring his pledge. Stationed there, he toes the Kremlin line against the West, harming Western democracies but not authoritarians.

Snowden's become a celebrity, but his disclosures have done little to ease the alleged surveillance he claimed plagued Americans. The data he stole from U.S. systems went far beyond information related to surveillance and included military secrets, putting Americans at risk and ruining costly protections paid for by taxpayers. Instead of reducing surveillance of Americans, Snowden increased surveillance of Russians. Two Russian journalists Andrei Soldatov and Irina Borogan masterfully detail how Putin has used Snowden and the threat of American spying as justification for mass internet surveillance in Russia in their book on the history of surveillance technologies in Russia. Snowden became a useful tool for passing more restrictive laws in that country, such as the "Bloggers Law" of May 2014, which "required bloggers with more than three thousand followers . . . to register with the government . . . [giving] the security services a way to track them, intimidate them, or close them down."[37] Russia has since banned VPNs that could hide someone's identity on the internet inside the country and has installed internet and social media surveillance systems that monitor all traffic in Russia. Snowden now rails against American cyber surveillance while living in a country with quite possibly the most intrusive internet surveillance in the world.

Mexican bloggers hung from bridges in 2011, and yet WikiLeaks has never mounted a campaign to avenge their deaths. There have been no calls to expose the violent tyranny of drug cartels and the corrupt government supporting them. No leaks from the Zetas cartel or the Mexican government have surfaced on the website. The Panama Papers and the Paradise Papers have spearheaded investigations of dark money, journalists have died, and WikiLeaks—well, they've gone a different route.

Through all of this, Russia has seen not a threat but an opportunity. Its government may have been on WikiLeaks' hit list, but it never feared transparency efforts, because it never would have tolerated it. No free press truly challenges Putin and his oligarchs. Russian journalists are either co-opted, jailed, or killed for challenging the government with secret disclosures. WikiLeaks' system from the beginning always favored authoritarians and harmed democracies, for the costs of disclosure in the West are at most detention, not death. Western bureaucratic agencies and political parties must have electronic communications, and the opportunities for a disgruntled employee to betray a nation, corporation, or elected official loom large. Those betraying our government know they'll never hang from a bridge or evaporate in a car bombing.

Instead, Russia intelligently recognized that transparency movements relied on content, and compromising information seeded to WikiLeaks provided a new method for character assassination. The Russian intelligence services had already forged ahead, compromising adversaries in cyberspace throughout the late 1990s and early 2000s. They secretly placed child pornography on the computers of defectors and intelligence officers and leaked salacious sex videos of political opponents onto the internet, creating a media feeding frenzy. Outlets like WikiLeaks were a perfect vehicle for widespread character assassination of enemies worldwide, an open-source vehicle for planting information that allowed for plausible deniability.

Many of the great chess masters have been Russian, and their leader, Vladimir Putin, is a lover of judo. Both require strategy, and victory routinely goes to those who employ their adversary's strengths against them. As Putin famously demonstrated his judo skills on You-Tube, Edward Snowden settled into a Kremlin-provided safe house. Julian Assange stowed away in the Ecuadorian embassy. The Kremlin trolls practiced on audiences in Ukraine and Syria, and occasionally heckled me. As for the hackers swirling around the Syrian Electronic Army, some of them went offline, busy working on a new project. And Russia's cyber team came together for a new mission, with some new methods the world had yet to see and still doesn't quite comprehend.

6

Putin's Plan

Soldiers surrender, sign armistices, and lay down their weapons when they're defeated. Intelligence officers—spies—burn files when their country loses, giving up *their* weapon: hard-won secret information on their enemies. That was exactly what Vladimir Putin did when the Berlin Wall fell, in 1989: he burned the Kremlin's files from his KGB outpost in Dresden, Germany. Putin's assignment outside Mother Russia focused on spotting, assessing, and recruiting East Germans with access to the West. He employed spycraft to buy, compromise, and coerce people into doing the Kremlin's bidding, focusing largely on "stealing Western technology and NATO secrets," in part by re-cruiting agents trained in "wireless communications." He also learned the subtler art of using politics and compromising situations, rather than overt force, to unseat and dethrone adversaries—a skill that proved handy during his rapid ascent to his country's helm upon re-turning home to the new Russia.

Putin saw the Soviet Union crumble as the West, aligned under the North Atlantic Treaty Organization (NATO), outspent and out-competed Communism on every level. The Soviets couldn't keep up with the American economy. Mikhail Gorbachev's move away from central planning and toward economic restructuring, known as per-estroika, combined with increased openness for political and social discussion—glasnost—came too late. Rather than adapt and upgrade Communism's competitiveness vis-à-vis the Western world, these liberalization efforts brought about the country's unraveling. Openly available information didn't free the Soviet economy; it crumbled it. Eastern European Communist dictatorships dissolved rapidly in 1989, marked by the epic fall of the Berlin Wall. Less than two years later, on Christmas Day 1991, the Russian flag replaced the Soviet flag atop the Kremlin.

Vladimir Putin understood relationships: how to mold them and manipulate them. Not surprisingly, his return to St. Petersburg led to his becoming a behind-the-scenes fixer for a controversial mayor before quickly climbing into Russia's leadership. In less than a decade, Putin rose from KGB veteran to president of Russia—a remarkable feat. He understood power: how to acquire it and how to wield it, even if armed with a weaker hand. Putin learned these skills in the KGB, where a weaker country, the Soviet Union, sought an asymmetric approach to undermining the West by nonmilitary means, an approach known as "active measures."

By the 1980s, the Soviet Union knew it was playing a losing game. American capitalism had exponentially outpaced Communism, providing the United States with significantly more military spending and economic power. Nuclear programs, combined with America's much vaunted new "Star Wars" missile defense system, put the Kremlin at a severe disadvantage at a time when the Soviets were experiencing their own calamitous war in Afghanistan. NATO further challenged the Soviet Union's grip on Eastern Europe. Short of the mutually assured destruction of a nuclear exchange, the Soviet Union would lose any major conflict with the United States and its broad range of allies. Economically, the USSR's closed planned economy was doomed to failure. The Soviets needed a new, cost-effective approach if they were to keep pace in the Cold War.

If the Soviet Union couldn't defeat the United States from the outside in, then it would have to collapse the United States from the inside out. Rather than fight a losing battle of military spending against overwhelming alliances, the Soviets accepted that the cards were stacked against them militarily, and developed a different approach to America and NATO.

"Active measures" became the tagline for the Soviet campaign to defeat the West "through the force of politics, rather than the politics of force." Ever since the United States created the United Nations, it had enjoyed dominance in the realm of "state-to-state" systems—diplomatic and military alliances. The Soviet Union, by contrast, created its alliances through the real or perceived threat of military force, establishing coalitions with iron-fisted strongmen in Eastern Europe and Central Asia. As the Cold War dragged on and the Soviets became increasingly isolated politically and militarily, active measures, led by the KGB's Service A, became the focal point for their efforts against America and the West.[1]

On the surface, the Soviets continued their state-to-state engagements and diplomatic contacts with the West, but they invested equal or even more energy in two other veins of influence. State-to-party efforts developed political parties inside Western democracies that promoted Communist agendas. This was a brilliant stroke, creating a double-edged sword to democracies promoting freedoms of speech, assembly, and press. If Communist ideals flourished and gained popular support, they might legitimately win in elections, leading to Communism's overtaking democracies from within. Conversely, American policymakers seeking to mitigate rising Communist sentiment would undermine their own democratic principles and tarnish American values, as witnessed during the 1950s period of McCarthyism.

Active measures also involved identifying individual citizens within Soviet adversaries who were ripe for influence and manipulation in the pursuit of Soviet objectives. Westerners driven by fame and fortune would be singled out and approached by KGB agents. Soviet strategists called these unwitting, malleable zealots "useful idiots," a term coined by Polish agents to describe Russian nihilists in the 1860s, who served as "useful fools and silly enthusiasts." If driven by ego, a good useful idiot provided a cheap method for influencing democratic societies at the grassroots level.

Soviet agents paired unwitting useful idiots with witting "fellow travelers"—Westerners supporting Communism and allies in the pursuit of Kremlin goals. Fellow travelers spoke the ideas, policies, and sympathies of Communism but resided in the West, where, both overtly and covertly, they were provided with political support or financial resources via Communist political parties. In tandem, these two kinds of targets could build a grassroots ground game, spreading the perception of organic support for Soviet ideas and objectives in Western democracies.

Soviet propaganda was the lifeblood of effective state-to-party and people-to-people active measures strategies. Soviets artfully blended three layers of messaging to influence their targeted populations. Their newspapers provided the baseline for Communist influence, spreading the Kremlin's party line, along with strategic falsehoods. Overt state-sponsored media outlets attributed to the Kremlin issued headlines known as "white" propaganda. "Gray" propaganda outlets were Soviet-established foreign newspapers, magazines, and radio and television stations. These semi-covert influence efforts amplified Soviet white propaganda, manipulated facts to spread exaggerated falsehoods, and fabricated local stories to smear Americans and promote Soviet policies and practices abroad. Local gray propaganda efforts, integrated with people-to-people strategies, created the appearance of indigenous authenticity.

Finally, there was "black" propaganda: covert actions by KGB agents to plant false stories, which appeared to the reader to come from a local source, and thus propel their legitimacy. Two key elements fueled these efforts: provocateurs and forgeries. To help spread falsehoods, KGB agents might order their assets to act as provocateurs, engaging in protests or crimes to add further credence to the information campaigns promoted by Communist newspapers all across the

West and spouted by Kremlin media outlets. Service A complemented these physical actions with legitimate-looking falsehoods. Bogus U.S. government internal memorandums, alleged letters attributed to the CIA, or bogus secret documents found their way to a wide range of news outlets carrying Soviet sympathies. During the height of active measures, the KGB crafted and strategically placed thousands of forgeries in media outlets around the world, soiling the American brand with conspiracies of all types. Forgeries and provocateurs leveraged the useful idiots and fellow travelers promoting Kremlin propaganda, thus providing local credibility to what might otherwise be obvious foreign meddling. The more local it seemed to be, the more successful an active measures campaign would be. When successfully employed by the Soviets, white, gray, and black propaganda dissemination provided a holistic information bubble, consuming targeted audiences across all media with synchronized, repeated messaging that would be difficult not to believe in the absence of a strong countereffort from the West.

Active measures propaganda didn't simply promote Soviet policy positions the way America would play to patriotism or democracy. U.S.-backed counters to Soviet propaganda, such as Voice of America, promoted freedom of speech, democratic governance, and free elections and hosted feel-good pieces on U.S. exceptionalism. The Soviet system, on the other hand, took a more negative, antagonistic approach, deploying a spectrum of messaging across four general themes. First, political messages, as one might expect, sought to tarnish the reputations of the Soviet Union's political adversaries or undermine democratic institutions by alleging corruption or incompetence. Next, social commentary messages played alongside political themes, fomenting racial, religious, and socioeconomic divisions among the American electorate. Financial propaganda, meanwhile, sought to undermine support for capitalism by stoking fears of world market collapse, wealth disparity, or imperialism. But above all, the Soviets attempted to inject fear into audiences. Fear, more than any other emotion, lowers people's ability to distinguish fact from fiction, making lies easier to sell. Audiences were reminded, relentlessly, of impending calamities that could bring the end of humankind. Nuclear standoffs with the West were amplified by Kremlin outlets at home and abroad. Global pandemics poised to destroy local communities offered a particularly effective line of attack on America, and this area of effort, more than any other, represents what may have been the greatest success of active measures to date.

The AIDS virus ravaged communities around the world, growing from a handful of cases in 1980 to more than 4.5 million in 1995.[2] The United States led much of the world's effort to counter the spread of HIV across impoverished regions. Instead of viewing the U.S. effort as a helping hand, though, much of the world then, and even now, believes that the U.S. government unleashed AIDS on the world as a biological weapon—thanks to the KGB's disinformation campaign known as Operation Infektion.

An active measures media campaign generally employs three simple ingredients to create damaging propaganda and provide the Kremlin with plausible deniability: anonymously sourced falsehoods, mixed with true information, disseminated through an information proxy. The Kremlin wanted to tarnish world opinion toward America as the Cold War ratcheted up militarily. It began one such campaign on July 17, 1983. An anonymous letter surfaced from an alleged "well-known American scientist and anthropologist" claiming that AIDS was the "result of the Pentagon's experiments to develop new and dangerous biological weapons." The letter stated that the United States planned to transfer experiments to Pakistan, a claim that would create panic for neighboring India, where the letter arrived. The *Patriot*, a little-known "gray" left-wing Indian newspaper partially founded by the KGB in 1967, received the letter and published a sensational story attempting to bolster the false anonymous claim with accurate facts about the AIDS epidemic, alongside public information about the U.S. biological weapons programs at Fort Detrick, Maryland. The

Patriot, a fringe outlet, received little attention from its false bombshell revelation against the United States.

While tensions between the United States and the Soviet Union rose, active measures picked up. On October 30, 1985, the *Patriot* AIDS conspiracy reemerged—this time cited by the KGB's overt propaganda newspaper *Literaturnaya Gazeta*. The article, entitled "Panic in the West or What is Hiding behind the Sensation Surrounding AIDS," pointed at the anonymous source in the *Patriot* letter from two years earlier. The story then detailed the biological weapons at Fort Detrick and the testing of LSD on CIA officers before alleging that the United States had conducted AIDS tests on unsuspecting victims. The KGB then directed its allied East German intelligence to further the conspiracy by adding more scientific detail to the narrative. They employed a "useful idiot," professor Jakob Segal, an agent known to the Soviets who authored a disinformation pamphlet entitled "AIDS: Its Nature and Origin." Segal provided extensive, detailed facts regarding the AIDS virus before falsely theorizing that the U.S. government had deliberately infected "homosexual prisoners who went on to infect gay populations in New York City and San Francisco."[3] Segal's pamphlet refuted the common consensus that AIDS had originated in Africa, instead asserting that America had invented the deadly virus. Segal proliferated this conspiracy during a presentation at the Eighth Conference of NonAligned Nations in Harare, Zimbabwe, September 1986.

Segal served as a Soviet proxy agent of influence, giving interviews to a wide variety of newspapers in West Germany and abroad, providing further legs to an unfounded conspiracy. By 1987, third world outlets routinely held the United States culpable for the spread of AIDS, repeating verbatim the manufactured Soviet falsehood. British newspapers, including the *Daily Telegraph*, repeated the claims, and the Soviets put special emphasis on countries hosting U.S. military bases, where "infected" U.S. soldiers might be spreading the deadly virus. Africans believed and repeated the false AIDS story the most, ultimately harming American efforts to combat the spread of the virus among the most affected populations. Even today, this Soviet-created conspiracy still endures in parts of the world. The Soviets orchestrated thousands of false stories and smear campaigns over the course of the Cold War. No single effort likely demonstrates the longevity and pervasiveness of active measures like Operation Infektion.

While individual campaigns like Operation Infektion achieved immeasurable results and a mix of intended and unintended consequences, the Soviet Union's active measures never materialized as a sufficient asymmetric counter for U.S. might and NATO's growth. Soviet propaganda outlets took many years or even decades to grow their audiences. Distributing messages and dollars to propel a Communist media insurgency in America required repetitive synchronization and significant resources in both manpower and production. Moreover, influencing populations in Western areas required layers of agents undertaking physical actions at the behest of the Kremlin. Exposure of Soviet operatives conducting active measures in the United States persistently jeopardized Kremlin foreign policy. Finally, American nationalism during the Cold War sustained a population averse to anything Soviet, resistant to Communist messaging and deeply suspicious of foreign influence. Stand-alone initiatives like Operation Infektion achieved remarkable tactical success, but strategically, active measures required too much time and money. They also required less resistance to cement themselves among targeted Western populations and to generate grassroots support. Active measures could and would work; the timing just wasn't right—until the advent of the internet.

\\\\\\\\\\\\\\\\\\\\\\\\\\\

Almost a year before Russia invaded Crimea in 2014, the chief of the general staff of the Russian Federation, General Valery Gerasimov, authored an article laying out his vision of future warfare based on his interpretation of recent Arab Spring protests across North Africa and the Middle East. Gerasimov noted:

> The very rules of war have changed. The role of non-military means of achieving political and strategic goals has grown, and, in many cases, they have exceeded the power of force of weapons in their effectiveness. . . . In North Africa, we witnessed the use of technologies for influencing state structures and the population with the help of information networks. It is necessary to perfect activities in the information space.[4]

Gerasimov asserted that Russia would be moving away from traditional ideas of conventional war, where battlefields defined the beginnings and ends of conflicts. Instead warfare would be conducted perpetually, on many fronts, with military action, particularly that of special operations forces, blended with political, economic, and, most important, information campaigns.

Only a few months after Gerasimov hinted at his military's future intentions, RIA Novosti News Agency disclosed the Russian Defense Ministry's formation of a separate branch of military forces aimed at combating cyber threats. The department evaluated candidates it wanted to work with, and Putin specifically noted that "so-called 'information attacks' are already being applied to solve problems of a military and political nature."[5] Active measures, something old, would be new again, this time using the advantages of the internet, cyberspace, and social media to accomplish what they could never do during the analog era of information warfare: dismantle democracies worldwide.

And we were watching—the "Jewish Asshole" Weisburd, the "Loud Shaytan" Berger, and I, "Big Watty Kafir." Similar to how Soviet intelligence had exploited race issues during the Cold War to divide American audiences, Russian influence efforts showcased violence and chaos all across the United States in the summer of 2015 as protests against police brutality broke out. Black Lives Matter demonstrations would be promoted and simultaneously scorned by the troll army, increasing distrust among the populace, law enforcement, and the government. Allegations of government misconduct might be seeded to agitate antigovernment groups. Government standoffs at the Bundy ranch, in Oregon, Jade Helm 15, and abortion protests all were showcased to fuel contempt among competing American factions. Traditional lines of active measures attack were all there on social media: political, social, financial, and calamitous. We considered writing up our analysis of the active measures renaissance, but we kept arriving at the same question: *Why?* In the fall of 2015, we didn't think Americans would understand Russia's active measures. Even if they did understand what was happening, I didn't think they would care.

The same could be said for the U.S. government. In the early summer of 2014, I provided a snapshot of the Russian social media campaign with regard to Syria as I closed a briefing on the Islamic State's rise.

"Have you all seen what the Russians are doing on social media?" I inquired.

The analysts were curious, but they were focused on counterterrorism, and ISIS's aggressive rise. Throughout the next year, I discussed Russian cyber influence whenever I had the chance during counterterrorism panels or government sessions, but the Islamic State's wave of violence suffocated any other impending threat. During a domestic extremism conference, another panelist studying antigovernment militias and white supremacists noted that she'd seen Russian influence pick up significantly in their online forums. At another security conference, a Russia national security expert remarked about how they were personally targeted by cyber attacks. And then there were rumors circling of hacks, big ones, many of them hitting American targets.

Russia's dedicated hacking campaign in the fall of 2015 proved to be like no other in history. Unlike the hacking tirades of criminals, Russia didn't pursue indiscriminate breaches for financial gain. It sought information from a select group—politicians, government officials, journalists, media personalities, and foreign policy experts—numbering in the thousands, according to government and media estimates.

Cyberattacks from Russia weren't new. The Kremlin had perpetrated cyberattacks as part of its military campaigns prior to invading Georgia in 2008, when it defaced and disabled Georgian government websites as part of a psychological warfare campaign. In 2014, a pro-Russian group called CyberBerkut surfaced alongside Kremlin hackers and penetrated Ukraine's Central Election Commission, altering the nationwide presidential vote in favor of Russia's preferred candidate, Dmytro Yarosh. Luckily, the Ukrainian government caught the manipulation before the results were aired. Despite this setback, the pace of Russian cyberattacks only quickened. Throughout 2015 and 2016, Ukrainian businesses and government agencies suffered endless cyber assaults. The most ominous Russian attack, known as BlackEnergy, struck the power grids of the Ivano-Frankivsk region of Ukraine, disabling electricity during one of the country's coldest periods, December 2015. These attacks, though, sought to damage infrastructure and undermine Eastern European countries through humiliation and confusion. The Russia-connected breaches surfacing in America, though, sought something different.

Putin's widespread hacks on America pursued privileged information about his country's Western adversaries. The stolen information the Russians wanted wasn't intellectual property, trade secrets, military plans, or bank account numbers, but rather compromising data on people, digital *kompromat* for discrediting reputations, sowing conspiracies, seeding false narratives, and ending careers. Hackers were gathering fuel for an active measures campaign like no other, an all-out operation to win the U.S. election.

Starting in the late summer of 2015 and extending through the fall, Russia undertook the largest, most sophisticated, most targeted hacking campaign in world history, breaking into the email accounts of thousands of American citizens and institutions. Analysts posit that the cyber offensive was perpetrated by two of Russia's intelligence agencies: the Main Intelligence Directorate, known by the acronym GRU, and the Federal Security Service, known by the acronym FSB, predominantly an internal intelligence arm but particularly sophisticated in cyber operations.

In cybersecurity speak, the GRU and the FSB operated as Advanced Persistent Threats (APTs), a reference to their dedicated targeting and wide array of cyber-hacking techniques. APTs, unlike common cybercriminals or hacker collectives, have sufficient resourcing to stay on their targets until they penetrate the systems they desire to access. APTs use a range of techniques, from the simple to the complex, employing all forms of social engineering and specifically tailored malware known as "zero days."

The Russian APTs were known in the cybersecurity world as APT28 (code name: Fancy Bear) and APT29 (Cozy Bear). Cozy and Fancy Bear represented competing Russian hacker groups seeking access and compromising information from democratically elected officials adversarial to Russia, media personalities (particularly reporters who interfaced with anonymous sources), military leaders, and academic researchers and policy think tanks studying Russia. In sum, anyone and everyone opposing Russia was targeted, in hopes that their private communications, if revealed, would undermine the credibility of a Russian adversary and/or sow divisions and mistrust between the targeted individual and those they maligned in private.

Common lore might suggest that Russia's hackers operate a complex system of hacking techniques and malicious code designed specifically to infiltrate American systems. But that's not really the case. Russia's hackers often use the most basic of techniques, relying on the underlying principles of social engineering to dupe unwitting computer users into coughing up log-in credentials to their email accounts, social media handles, and websites.

"Spearphishing" remains the most useful and most common mechanism for gaining access to users' accounts. Every internet user has encountered spam tempting them to click on a link for an amazing deal at a favorite store or claim a prize they've won. But Russian spearphishing focused more squarely on injecting fear of a breach to actually achieve a breach. Many of Russia's targets received what appeared to be legitimate warnings to reset their email passwords. These spearphishing emails trick unsuspecting users into clicking on a link that redirects them to what's known as a "watering hole" website—a site that appears to be a well-known legitimate portal but is actually a fake page requesting a username and password. Common watering hole attacks mirror the home log-in pages of banks, email providers, social media platforms, and student portals. The user unwittingly enters his or her username and password on the watering hole, and instantly hackers gain the target's log-in criteria. Hackers then use that username and password to access the true email of the target and download their private communications.

The Kremlin election hacking wave began in the fall of 2015. We all remember the most critical and ultimately damaging hack—when the Democratic National Committee was breached. In September 2015, a D.C.-based FBI agent notified the DNC's tech support contractor Yared Tamene via a voicemail, regarding a potential intrusion at the DNC. But Tamene didn't react to the notifications of Special Agent Adrian Hawkins, who in recent years had been tracking a Russian cyber-espionage group called "the Dukes". Both Fancy Bear and Cozy Bear breached the DNC in separate attacks, roaming the party's computers for seven months, stealing emails, communications, and records—a treasure trove of information ripe for *kompromat*. Separately, hackers penetrated the Democratic Congressional Campaign Committee sometime around March or April 2016. Hackers also hit the Republican National Committee (RNC), but the GOP got lucky, compared with its political rivals. That intrusion was smaller and struck an old RNC server no longer in use, rendering virtually no compromising materials. Some sources suggested that the old server had previously been used by Senators Lindsey Graham and John McCain—two well-known adversaries of Russia.[6]

By the start of 2016, Russia had gone from spearphishing of political parties to "whalephishing" of key political operatives and government officials. Whereas spearphishing targets swaths of accounts, seeking many entry points and access to unknown data, whalephishing targets prominent individuals inside organizations or governments whose private communications likely provide a wealth of insight and troves of secrets to propel conspiracies. John Podesta, campaign manager to Hillary Clinton, proved to be the biggest whale hacked in 2016.

Podesta received an email that appeared to be from Google, alerting him of an improper attempt to log in to his Gmail account. The message, designed by social engineers to inject fear of compromise into the target's mind, redirected the user to change his password by clicking a button in the email. Charles Delavan, a Clinton campaign aide notified about the warning and tasked with checking the message's legitimacy, claims he made a typo in his response.

"This is a legitimate email," Delavan messaged to one of Mr. Podesta's aides. Later Delavan would say he'd meant to type "illegitimate" and that his typo implied the opposite of what he intended.[7] The link in the whalephishing email was clicked, and very soon about sixty thousand emails had been taken from Podesta.[8] Retired flag officers in the military, both current and former, encountered the same scheme. Former secretary of state and chairman of the Joint Chiefs of Staff Colin Powell lost control of his account, as did a former commander of NATO, General Philip Breedlove. Washington, D.C.'s academic think tanks that had programs focusing on Russia, if they didn't detect it on their own, received warnings from the government or from cybersecurity companies like CrowdStrike.[9] Post-election reports revealed that Russia had issued expertly crafted Twitter messages to more than ten thousand U.S. Department of Defense users. The malware enabled Moscow-based hackers to take control of the victim's phone or computer. The Kremlin left no target untouched.[10]

Meanwhile, the troll army's interest in the U.S. presidential election gained steam toward the end of 2015. One article in particular caught my eye.

"Is Donald Trump a Man to Mend US Relations with Russia?" Sputnik asked on August 24, 2015.[11] Trump's campaign, at the time, seemed more celebrity stunt than deliberate effort to lead the nation, but the post was curious, given that Russian disdain for both parties and their leaders had historically been a constant.

From then on, the social media war in America surrounding the election proved unprecedented, and the Russians were there and laying the groundwork for their information nuclear strike. Russian state-sponsored media, the English-speaking type, was quite clear: Putin did not want Hillary Clinton to become president. Aggressive anti-Clinton rhetoric from state-sponsored outlets, amplified by their social media trolls, framed Clinton as a globalist, pushing democratic agendas against Russia—an aggressor who could possibly bring about war between the two countries. The trolls' anti-Clinton drumbeat increased each month toward the end of 2015 and going into 2016. The Kremlin spotted a new, more likable alternative among the Democrats, Bernie Sanders, whose challenge to Clinton was growing each day and whose message rang with socialist themes. Meanwhile, Trump's brash barbs against his opponents were working unexpectedly well. Kicking off 2016, the troll army began promoting candidate Donald Trump with increasing intensity, so much so their computational propaganda began to distort organic support for Trump, making his social media appeal appear larger than it truly was.

Russian leaders, much like their boastful American counterparts, have egos and often can't help themselves when they want to brag.

That appears to be the case with Andrey Krutskikh. Speaking at In-foforum 2016, in Moscow, he hinted at the Kremlin's plans. The Russians were implementing new strategies for the "information arena," he said, echoing General Gerasimov's doctrine from a couple of years earlier. Krutskikh compared deployment of Moscow's new influence weapons to that of a nuclear bomb that would allow the Russians "to talk to the Americans as equals."[12]

Curiously, on March 16, 2016, during the height of the primary season, WikiLeaks launched a new database that provided users with a search function for moving through and identifying topics among more than fifty thousand pages of emails from Clinton's private server. The State Department had previously released some of the data, but WikiLeaks took the added step of creating a rapid system for accessing these emails around key words. The archive provided a novel feature for Clinton's opponents seeking to gather dirt on the Democratic candidate and greatly assisted journalists looking for juicy campaign stories. All the while, those attending Trump rallies screamed about 33,000 missing emails and calling for her to be "locked up" for disclosing classified information—referring to the emails deleted from the private server of Hillary Clinton that were unaccounted for in the FBI investigation. But on July 5, FBI director James Comey concluded his investigation into Secretary Clinton's emails and recommended no formal charges. The chants would continue, and the searchable database of WikiLeaks would take on a new life.

The fuel for Russia's new cyber active measures appeared on July 22, 2016. The twenty thousand emails and more than eight thousand attachments stolen from the DNC surfaced on WikiLeaks. The compromising information covered internal communications from January 2015 to May 2016 and was made available to the public just three days prior to the Democratic National Convention. Media coverage of the convention became distracted by conflict and conspiracies. The emails pointed to DNC suppression of the Bernie Sanders campaign, creating a third theme that Russian troll networks reinforced: that the Democratic Party was corrupt and Bernie Sanders got a raw deal, never having a chance to defeat Hillary Clinton. Revelation of DNC chairwoman Debbie Wasserman Schultz's private remarks showing her favoring Clinton over Sanders led to her resignation, and the mainstream media ran wild with the leaked information. The Russian leaks tarnishing Clinton worked, and they were just beginning.

Five days after WikiLeaks' dump of DNC emails, Donald J. Trump took to the stage at a press conference in Doral, Florida, and announced, "Russia, if you're listening, I hope you're able to find the thirty thousand emails that are missing . . . I think you will probably be rewarded mightily by our press."[13]

I watched the clip several times, and a sick feeling settled in my stomach. I'd watched the Russian system push for Trump and tear down Clinton, but up to that point, I hadn't believed the Trump campaign might be working with the Russians to win the presidency. I'd given briefs on the Russian active measures system in many government briefings, academic conferences, and think tank sessions for more than a year. But nothing seemed to register. Americans just weren't interested; all national security discussions focused narrowly on the Islamic State's recent wave of terrorist attacks in Europe. I did what most Americans do when frustrated by politics: I suffered a Facebook meltdown, noting my disbelief that a U.S. presidential candidate would call on a foreign country, one already pushing for his victory, to target and discredit a former first lady, U.S. senator, and secretary of state.

The U.S. government wasn't interested in the research; Americans weren't interested in the analysis; I could find no way to turn the Russia research from hobby to paid work. What was I going to do with this massive waste of time and research? If WikiLeaks was going to dump stolen information, then I would start dumping research onto the internet. I wouldn't have the impact of WikiLeaks, but at least I'd try.

On Saturday night, July 30, just three days after Trump's call for

Russian assistance, two stories were released in short order by RT and Sputnik. The state-sponsored news agencies both published articles about possible protests at the U.S. air base in Incirlik, Turkey, an increase in Turkish security forces, the potential for a second Turkish military coup, and leaving U.S. nuclear weapons in jeopardy. Minutes later, key Twitter accounts that Andrew, J. M., and I had watched for some time began tweeting both stories. "We have a situation in #Turkey #Incirlik" was followed by sparks and bursts of tweets raising alarm bells about a potential attack at the base. One group of accounts panicked about the possibility of nuclear weapons getting lost to hordes storming the base. A separate group compared the impending doom to Benghazi, and then another subset of tweets asked why the mainstream media wasn't responding to the incident. Amid these groups, bots promoted hashtags amplifying four themes, at distinct time intervals. #Media #Trump #Nuclear #Benghazi—the pumped hashtags painted the intent of the campaign. The media wasn't reacting to the incident at Incirlik, though, because nothing was happening. The small protest at the gate had been there for some time and seemed not at all threatening. Preparations for the U.S. chairman of the Joint Chiefs landing the next day provided a reasonable explanation for the added security at the base. On the ground in Turkey, nuclear weapons were not falling into the hands of any military coup or protesters, but in the social media world, one might have thought a second Benghazi-style attack was about to occur.

Analysis of the Twitter accounts showed a familiar pattern. The first ninety minutes of tweets using the #Incirlik hashtag circled around known amplifiers we'd previously observed. The account bios used highly similar words, most commonly "God," "country," "family," "conservative," "Christian," "America," "Constitution," and "military." While there certainly might have been real Trump supporters watching and praying for the safety of U.S. service members on that Saturday night, most of those inciting panic about the Incirlik attack weren't real accounts; they were social bots creating the appearance of being Trump supporters, hoping to propagate strategically placed false narratives among real Trump supporters.

Weisburd and I published the analysis of the Incirlik social media campaign at the *Daily Beast* on August 6, 2016,[14] and the story received a good number of views and some commentary. But the article's release quickly succumbed to the endless barrage of Trump–Clinton election coverage. Despite Russia's efforts, Trump looked to be a long way from victory, his poll numbers sliding as Clinton took what appeared to be a commanding lead. I resumed my day job, but once again, news coverage of the Trump campaign incited fear.

"You had the NATO base in Turkey being under attack by terrorists," Paul Manafort exclaimed to Jake Tapper on CNN on August 14, 2016. The Trump campaign again appeared to be relying on Russian information operations to power their narratives. For the second time in a month, I wondered whether the campaign might be actively working with Russia to influence Americans. The Incirlik coup—or attack, or whatever it had been depicted as—had been debunked almost immediately, and now Trump's campaign manager was using it as a talking point to discredit mainstream media and suggest bias against the Trump campaign. The next day, the *New York Times* reported that Manafort's name had appeared on a payment list belonging to Russia-backed Ukrainian president Viktor Yanukovych. On August 19, shortly after the Incirlik claims and the revelations of the Ukrainian ledger, Paul Manafort's leadership of the Trump campaign ended.

Amid the leaks and campaign battles, an online persona named Guccifer 2.0 emerged in June 2016, claiming credit for delivering hacked DNC emails to WikiLeaks. Analysis from multiple cybersecurity research companies, however, pointed to APT28 and APT29 connections to the DNC hack, not this new persona Guccifer. Guccifer 2.0 claimed to be a Romanian hacker, but technical signatures from Guccifer emails pointed to a predominantly Russian-used VPN and limited Romanian language ability. All signs suggested that Guccifer 2.0 was a cutout for the Russian government, providing the

Kremlin with plausible deniability of any involvement and a conduit for future releases. Guccifer did just that on August 15, releasing hacked materials from the Democratic Congressional Campaign Committee that discussed congressional campaigns and candidates' strengths and weaknesses.

Roger Stone, a Trump adviser who'd appeared months before on RT accusing President Clinton of sexual abuse, spoke to a crowd in Florida on August 10, claiming that he'd communicated with Julian Assange of WikiLeaks. Stone publicly promoted Guccifer 2.0 as the hacker behind the DNC breach, calling Guccifer a "HERO" and privately sending direct messages via Twitter to the @GUCCIFER_2 Twitter account between August 14 and September 9, 2016.[15] On August 21, he tweeted, "It will soon the [sic] Podesta's time in the barrel #CrookedHillary," foreshadowing an upcoming round of hacked emails.[16] Again I watched with concern, trying to assess the linkages between Russian hacking and influence alongside the peculiar connections, actions, and words of the Trump campaign.

Donald Trump's chances of being elected appeared to have been dashed when the *Washington Post* published the transcript of a previously unknown video in which he bragged about his dominance over women, commenting that he could "grab 'em by the pussy." Less than an hour after this video surfaced, though, WikiLeaks published thousands of the emails stolen from Hillary Clinton campaign chairman John Podesta in March. The emails showcased paid speeches Clinton gave on Wall Street, and their release could further the rift between her and the working-class voters she hoped to court. In a second release, on October 11, 2016, WikiLeaks disclosed that Donna Brazile, a CNN spokesperson, had sent messages regarding debate questions to the Clinton campaign. CNN parted ways with Brazile shortly after, chalking up another casualty of Russian active measures. From July through the opening of the polls, WikiLeaks dropped dribs and drabs of stolen information to continue powering the social media storm against Clinton.

Russia's pushing of the Trump train continued all the way up to Election Day, but by October the Kremlin's messaging turned a bit, suggesting that they, too, had read the polls and were possibly anticipating a Clinton victory. Beyond tearing down Clinton, pumping Trump, and noting that Bernie Sanders had gotten a raw deal from the Democratic Party, Russian state-sponsored outlets and their troll army gave increasing coverage to two allegations echoed by candidate Trump. Claims of voter fraud took on a new intensity, despite no evidence of false voter enrollments. Second, the Kremlin pushed that the U.S. election was rigged for a Hillary Clinton victory, suggesting that the vote count wouldn't be true. Ominously, Trump took to the stage repeatedly in October and November voicing the same conspiracies, each time offering no evidence to support his claims.

Most Americans probably saw Trump's claims as a sign of a sore loser preparing for defeat, but Russia's backup plan to undermine American democracy through active measures benefited from his allegations. The first wave of Russian hacking, starting back in the fall of 2015, had sought compromising information on those tied to the Clinton campaign and the presidential election in general. The second Russian hacker wave, in 2016, aimed to sow chaos on Election Day. The Department of Homeland Security began receiving calls from state election bodies complaining of hacks on their systems. In total, Russia-connected cyber actors attacked some part of the voting infrastructure in twenty-one states. The goal wasn't to change votes, necessarily, but instead to tamper with state and local voter rolls.[17] Previously registered voters might show up to the polls and their names would be missing from the list, or false names would be added to the rolls, which would then provide ammunition to myths of voter fraud. Should Hillary Clinton be elected, it would be much easier for Russia to use an influence campaign to undermine her mandate to govern as conspiracy questioned the legitimacy of her victory.

As active measures pushed Putin's preferred agenda in America and at home, the troll army grew in size and intensity. Each WikiLeaks release of stolen emails not only powered Kremlin state-sponsored outlets but also armed conspiracies from fringe media sites and political opportunists. Clickbait websites in Macedonia and Canada emerged from thin air, pushing false election-related news stories that lured viewers and advertising revenues. Political parties and the super PACs supporting them launched sustained media campaigns, using social media to amplify their message. Members of the Trump campaign, in particular retired lieutenant general Michael Flynn, cheered "Lock her up!" on campaign stages and amplified WikiLeaks' links on social media. Russian state-sponsored news promoted Trump conspiracy theories of voter fraud and election rigging while WikiLeaks disclosures of hacked material powered wide-ranging smears of Clinton. With the DNC fractured and in disarray, Russia hoped that Bernie Sanders supporters would stay home on Election Day.

The cacophony of social media rage grew in the final two months of the campaign as Trump and Clinton met for three debates. Political social bots picked up steam. The first debate witnessed massive surges of Twitter activity. By the second debate, researchers at Oxford University determined that nearly a fifth of pro-Clinton and a third of pro-Trump tweets originated from automated accounts. The third debate, on October 19, showed pro-Trump bots outnumbering pro-Clinton bots by a margin of seven to one.[18] Trump supporters joined in, proclaiming victory and outshouting Clinton supporters online, while in reality, commentators mostly called the outcome a toss-up or gave a slight edge to Clinton.

It appeared that Putin's active measures and open support for Trump wouldn't be enough to overcome the nearly endless string of Trump gaffes and less than stellar debate performances. Like many, I nervously watched the polls going into the final weeks. While I found little that excited me about the Clinton campaign, Trump's overt love for authoritarians like Putin and parroting of Kremlin propaganda made me incredibly uneasy. Pundits seemed convinced after the final presidential debate that Clinton would win easily. I became more convinced that if Clinton won, the Kremlin would only intensify its influence efforts on the American public, which made me wonder whether, after the election, I'd finally be able to gain interest from the U.S. government for studying Russian online active measures.

My thinking changed quickly on the afternoon of October 28, 2016, when FBI director James Comey sent a letter to Congress reopening the investigation into candidate Clinton's emails from the State Department. Former congressman Anthony Weiner, embroiled in an investigation regarding his explicit conversations with an underage girl, had surrendered a computer to investigators that contained communications between his wife, Huma Abedin, and then–Secretary of State Hillary Clinton. All of Trump's gaffes, bad debates, and lack of policy expertise quickly evaporated, and the election changed overnight.

The single biggest issue for pro-Trump online propagandists, Russian or otherwise, was missing emails, and Director Comey's letter ignited a firestorm on Twitter. The following weekend saw nonstop coverage of the Clinton email controversy—an American couldn't engage with any media without hearing of the renewed investigation. Almost immediately, polls swung in the opposite direction and states marked solid blue on campaign tracking maps swooped back into play. Having conducted opinion polling during the war on terror, I watched FiveThirtyEight's tracker cautiously, and the volatility in the previous months convinced me that no one had any idea how people would actually vote on Election Day. The monthly polling swings— Clinton up by five and then neck and neck with Trump—made it clear to me that the election was impossible to call. On Halloween night, one week before Election Day, I checked FiveThirtyEight's polls one more time—they were basically showing 50–50, an even split, a complete toss-up.

I had a decision to make. If Trump won, there would be no future Russian active measures research by the U.S. government; I felt certain of that. Although almost anything seemed plausible at this point,

Trump seemed more of a "useful idiot," opportunistically taking the Kremlin's help, than a "Manchurian candidate" infiltrating the U.S. political system under Russia's direction. Either way, his overtly pro-Russian foreign policy stances and adoration of Putin convinced me that a future Trump administration wouldn't be working aggressively to counter the foreign meddling in this election. I'd also learned on my Facebook and Twitter feeds that my questioning of Trump's Russia ties led conservatives to label me a liberal shill for Hillary Clinton. Despite my upbringing in red-leaning Missouri and my reputation as a pro-drone war hawk during the Obama years, Republican friends suddenly saw me as a Democratic operative seeking to tear down Trump. If I waited to publish research pointing to Russian interference and influence on behalf of Trump until after he won the election, I felt it would be seen as some liberal snowflake conspiracy theory. The more I pondered a potential Trump win, the more certain I became that our findings on Russia's renewed active measures in cyberspace must be published right then, before the election.

I called Berger and Weisburd. I urged them to join in writing up our findings, explaining why I thought we had to get our work out before the election. Again, we all felt some reluctance about penning another unfunded article for which we'd likely suffer renewed cyber-attacks.

I started writing with less than a week till Election Day, and Weisburd and Berger joined me. For three days, we put together our final accounting of what we'd observed in the nearly two and a half years since we stumbled onto the Kremlin's trolls. We edited and deliberated late into Friday night before the election, and I sent the final copy to War on the Rocks, a rising national security blog we'd written for in the past. I offered it to them first, under the stipulation that they release it at least two days before the election, before any of the late-breaking drama likely to come in the final hours of the two campaigns. On Sunday, November 6, 2016, "Trolling for Trump: How

Russia Is Trying to Destroy Our Democracy" posted at the War on the Rocks website, and we braced for future cyberattacks. My colleagues and I sent the article around, and in foreign policy circles we received a good number of reads and feedback, but the cacophony of the election buildup quickly buried our analysis and the three of us returned to our day jobs. And, like most everyone else, all we could do was watch election coverage.

By Election Day, allegations of voter fraud and the election being rigged created such anxiety that I worried that some antigovernment and domestic extremist groups might undertake violence if Trump didn't win. Pundits continued their prognostications of an easy Clinton victory, and calls for Trump supporters to go to the polls and monitor for fraud echoed far and wide on social media. Russian social media influence networks prepared for a Trump loss and promoted stories about the election being rigged as much as they pumped candidate Trump.

Worn out from the unending election cycle and years of Russia analysis, I cast my vote in upstate New York and headed to an election-watching show at the Comedy Cellar, in New York City, my favorite place to hang and a spot I hoped might provide some needed relief after so many stressful months.

The polls and the pundits who read them seemed convinced that Hillary Clinton had survived the ten days of turmoil created by the Comey letter, since the FBI director had announced that the newly discovered emails had yielded no further evidence in the Clinton investigation and the case was again closed. I kept a close eye on my Twitter feed throughout the day, looking for signs of conflicts or violence at polling places, praying no armed individuals would surface and do damage based on fake news. Russia's trolls promoted the #voterfraud conspiracy at a steady pace throughout the day, hoping to sow chaos and create doubt.

The first round of results came in as expected: Pennsylvania and

Florida looked too close to call. Then came the 8 p.m. results, and the comedians all but trumpeting a Clinton victory fell quiet. I looked at the real-time election board: Michigan was showing for Trump.

Around the room, the results weren't yet sinking in. I walked outside to get a better signal on my phone, and sure enough, Trump was winning. I went back inside and the comedians were struggling to keep the laughs going. The audience was confused, distraught, but, while I was surprised Trump was winning, I *could* believe it. The polls had been haywire for months, and only the actual, real vote showed voters' true sympathies. I checked in with my colleagues, and the themes of voter fraud and election rigging were waning on Twitter. The Russians were just as surprised as we were, I imagined. And now they had their own problem: in hoping to undermine Clinton through conspiracies of election rigging, they now risked undermining their preferred man, Trump.

By 11 p.m., Trump and Clinton supporters were both in disbelief. Trump was going to win. It was only a matter of time before the results were finalized. I smoked a cigarette on the street in front of the Comedy Cellar, something I've rarely done in my life, probably not since I partied on Bleecker Street during my college years. I watched distressed New Yorkers make their way home. I wondered: *Are they popping champagne bottles and clinking vodka shots in the Kremlin right now?* Putin's plan had worked: an American reality show star spouting Russian propaganda lines would soon be the president of the United States.

7

Postmortem

The postmortem on Russia's influence and meddling in the presidential election of 2016 began well over a year ago and may never end. Less than a month after the election, social media influence became a fixation with journalists as they looked for an explanation as to how Donald Trump had beaten all the odds. He was completely unconventional, uninformed, unlikable in so many ways, and yet he had become the leader of the free world. *Fake news* entered the American lexicon, and my team's pre-election detailing of Russian active measures on the internet was now the subject of hot debate. Had fake news swayed the U.S. presidential election? A *Washington Post* article cited our study, and soon left-leaning trolls, led by the self-righteous Glenn Greenwald, of the *Intercept*, and the always bitter Matt Taibbi, of *Rolling Stone*, grouped me with McCarthyites seeking to suppress free speech. Right-wing fanatics, aided by Russia's trolls, presented me as a Clinton apologist, a sore loser always out to get President-elect Trump. I'm certain I'm neither and also can't be both at the same time.

Beyond the public back-and-forth, social media companies began digging into the data, and what they found spelled dangerous trends for democracy. Americans were increasingly getting their news and information from social media rather than mainstream media. Users weren't consuming factual content. Fake news—false or misleading stories from outlets of uncertain credibility—was being read far more than that from traditional newsrooms. *BuzzFeed News* analysis of the final three months of the campaign showed that Facebook users accessed false news stories at rates higher than mainstream news. *EndTheFed.com* and *Political Insider* produced four of the five most read false news stories in the three months leading up to the election. One story famously and falsely claimed that Pope Francis had endorsed Donald Trump and another story claimed that Hillary Clinton's emails hosted on WikiLeaks certified her as an ISIS supporter. Three of the five most read articles, though false, promoted and referenced Russia's hacking of Americans and delivery of stolen contents to WikiLeaks.[1] Throughout December, fears of Russian election manipulation grew, and each day brought more inquiries into how Russia had trolled for Trump.

"There's no evidence of collusion" had become a constant scream from Trump supporters by the summer of 2017. Trump opponents, self-labeled on Twitter as the #Resistance, saw conspiracy and Russian collusion at every turn, hyperventilating with each new revelation connecting Team Trump with Russia. Months of Russia investigation by Congress, the firing of FBI director Comey, and the appointment of Special Counsel Robert Mueller, a former FBI director, to examine Russia's interference in the election have extended Vladimir Putin's victory over the United States. The American electorate remains divided, government operations are severely disrupted, and faith in elected leaders continues to fall. Americans still don't grasp the information war Russia perpetrated against the West, why it works, and why it continues.

The Russians didn't have to hack election machines; they hacked American minds. The Kremlin didn't change votes; it won them, helping tear down its less-preferred candidate, Hillary Clinton, to promote one who shares their worldview, Donald Trump. Russia's ability to influence the 2016 vote comes from many structural factors in American democracy. Narrow bipolar races make foreign influence such as that by the Russians particularly easy and effective. In the cases of both the United Kingdom's 2016 Brexit vote and the U.S. presidential election, the difference between the two sides amounted to merely one percent. A slight nudge at the polls provided by strategically leaked *kompromat* can send a preferred candidate over the top or suppress turnout for supporters of Kremlin foes. In two of the past five elections, the candidate who lost the majority vote—Bush in 2000 and Trump in 2016—ascended to the presidency. Electoral College disenfranchisement gives outside influencers the ability to enter into an electoral space and strategically target voters in only a few states with propaganda. From a laptop, anyone can microtarget a state's or county's voters on social media through deliberate analysis. The duration of U.S. presidential campaigns also assists adversary manipulation. Trump's rise occurred over almost sixteen months, stretching from the summer of 2015 to the end of 2016. Long primary and election campaign seasons provide extended periods for Russian propagandists to infiltrate key audiences and then direct them on strategic themes and messages supportive of Kremlin-preferred candidates.

Americans' rapid social media consumption of news creates a national vulnerability for foreign influence. Even further, the percentage of American adults fifty and older utilizing social media sites is one of the highest in the world, at 50 percent. Younger Americans, aged eighteen to thirty-four, sustain a utilization rate above 80 percent.[2] Deeper analysis by the Pew Research Center shows that U.S. online news consumers still get their information from news organizations more than from their friends, but they believe that the friends they stay in touch with on social media applications provide information that is just as relevant. Online news consumption moves increasingly each year from desktops and laptops to mobile devices; mobile has seen more than 20 percent growth in just three years. Social media apps such as Facebook dominate user time on mobile devices, pointing to the natural shift of news consumption to these applications.

The American move to online and mobile has been accompanied by a massive growth in online news outlets that suit the preferences and biases of Americans. Diffusion enables Russian active measures to infiltrate American audiences through multiple strategic placements narrowly targeted to audiences that can tip an election.

The *Columbia Journalism Review* analyzed the news outlets most frequently shared by supporters of Trump and Clinton. Fans of both candidates demonstrated a proclivity for outlets supporting their political biases, but the differences between the two camps were stark. On social media, Clinton supporters shared the *Washington Post*, *Huffington Post*, and *New York Times* the most. Trump supporters far and away preferred *Breitbart*, the *Hill*, and Fox News. Further down the list, Clinton supporters gravitated to a wide range of liberal outlets, most fairly well known. Trump's camp, though, included a long list of lesser-known outlets, including the controversial *Infowars*, a media organization known for denying the occurrence of the Sandy Hook shootings and one I'd witnessed routinely regurgitating Russian propaganda.

A cursory look at the *Columbia Journalism Review*'s media map demonstrates how social media encourages information bubbles for each political leaning. Conservatives strongly centered their consumption around *Breitbart* and Fox News, while liberals relied on a more diverse spread of left-leaning outlets. For a foreign influence operation like the one the Russians ran against the United States, the highly concentrated right-wing social media landscape is an immediate, ripe target for injecting themes and messages. The American left shows to be multipolar, littered with fringe outlets and causes, making concentrated foreign influence more challenging; spreading the Kremlin message thus requires influencing many outlets rather than one or two. Above all, the study shows how damaging Clinton's emails were to the campaign. Even among mainstream news outlets, Clinton's emails, a topic that the Kremlin pushed massively, was the subject of more than 65,000 sentences—nearly double any other campaign issue and far outweighing any Trump scandal.[3]

Our pre-election article on Russian social media active measures placed me squarely in discussions regarding whether President Trump legitimately won the election or if Russia did it for him. These debates rage on today, well into Special Counsel Robert Mueller's investigation and multiple Senate and House intelligence committee investigations into Russian meddling. *Did Russia win the election? Did Russia collude with the Trump campaign?*

These questions are not mutually exclusive. They overlap and diverge, and the biggest ones will prove vexing and, ultimately, indeterminate.

In my opinion, Russia absolutely influenced the U.S. presidential election of 2016. The single largest theme echoed throughout the campaign was Hillary Clinton's emails. Russia's theft and repeated release of emails from the DNC and the Clinton team powered and sustained a narrative of corruption, criminality, and conspiracy that clouded the Clinton campaign from start to finish. Trump and his advisers directly cited WikiLeaks releases and opportunistically repeated Russian narratives. Chants of "Lock her up!" and "emails!" still—more than a year after the election—ring out at Trump rallies. Some have suggested that political campaign propagandists were the ones using social bots to influence Trump supporters, but these claims fail to account for the value of the material that Russia obtained through hacking, which powered those political attacks against Clinton. Same goes for fake news peddlers in Macedonia and other locales who used sensational headlines and stories to create clickbait for advertising revenue. Of course, they influenced the election, but their bogus narratives benefited from the one thing Russia did that no one else could do: strategically hack and release *kompromat*.

Russia didn't need anyone's help to target Americans online; instead, it helped everyone else by hacking and releasing confidential secrets. Some conspiracies claim that Russian social media influence efforts worked in concert with the Trump campaign to microtarget specific portions of the electorate. I've seen no evidence to support this theory, and while I do believe that political groups benefited from Russian influence, that doesn't necessarily mean that those political groups knowingly colluded with Russia in cyber influence operations. I observed Russian social media operations dating back to 2014. Many of these accounts ultimately turned toward U.S. audiences in 2015 and later informed online discussions of the presidential election in 2016, but they predated the Trump campaign. Political social media advertising can easily and quickly repeat effective social media manipulation from Russia or other influencers without directly coordinating their efforts.

Russia alone did not win the election for Trump. It certainly helped the race remain close at times when Trump might have fallen completely out of the running. A prime example of its assistance is the strategic dumping of the stolen John Podesta emails less than an hour after Trump's disastrous, sexist "grab 'em by the pussy" comments hit the airwaves. Efforts like this offset attention from potentially catastrophic incidents in the Trump campaign. But Russian support for Trump and derision of Clinton can't be separated from the FBI investigation of then–Secretary of State Hillary Clinton's email server, the July 5 announcement by FBI director Comey closing the case, and the monumental shift arising from Director Comey's announcement, one week before the election, that new emails had surfaced, requiring a reopening of the investigation.

Clinton, Putin, and Trump were all equally shocked that Trump won the election. Without the Comey letter, I believe Clinton would have won the election. Russian influence networks also shifted their themes to voter fraud and election rigging in preparation for a Clinton win, as a way to undermine her mandate to govern should she lose. The dossier compiled by ex-MI6 agent Christopher Steele on

Trump, and a study published by the Russian Institute for Strategic Studies, a Moscow-based think tank established by Boris Yeltsin in 1992, both pointed to this shift in strategy by October 2016.[4] Without the Russian influence effort, I believe Trump would not have even been within striking distance of Clinton on Election Day. Russian influence, the Clinton email investigation, and luck brought Trump a victory—all of these forces combined.

Facebook, Google, and Twitter are likely the only ones with sufficient data to determine whether Russian influence won the election for Trump. They could track the spread of specific Russian-powered themes to specific geographies in the United States. Clicks in key swing states could be calculated and compared with political polling to see how certain themes or messages emanating from the Kremlin may have influenced perceptions. The equation would be fairly straightforward: How did false news and manipulated truths from Kremlin influence networks shift votes from Democrats and independents to Trump or bring about decreased voter turnout among Democrats in swing states?

Even if this electoral arithmetic were possible, one metric remains elusive: accurate political polling going into Election Day. Almost every poll leading up to the U.S. presidential election was wrong, and pollsters missed the mark repeatedly during the 2016 campaign year. Voter shifts due to Russian hacks and influence operations will be impossible to accurately account for, as there is no true measure for public opinion. No one will ever be able to prove without a doubt whether Russia did or did not win the election for Donald Trump.

\\\\\\\\\\\\\\\\\\\\\\\\\\\\\\

Having watched Russian influence leading up to the election, my hypothesis is that Putin won at least two states for Trump: Michigan and Wisconsin. Here's why. The first step when analyzing influence is to look at the audience's media usage. Of the key swing states in the 2016 presidential election, Wisconsin and Michigan have slightly higher internet penetration and mobile usage than the others, so there's a greater chance that voters there were exposed to Kremlin influence.[5] Next, Michigan and Wisconsin proved to be the closest races in the election, decided by less than 1 percent of the votes. Only 10,700 votes in Michigan and 22,700 votes in Wisconsin separated the two candidates.[6] Prior to the general election, Hillary Clinton struggled in these two states, losing both primaries to Bernie Sanders. These losses made Michigan and Wisconsin voters ripe for all three of the principal themes Russia pushed leading up to the election: Clinton's emails, her corruption and potentially poor health, and narratives of Bernie Sanders getting a raw deal from the Democratic National Committee. A minor theme pushed by Russia's social media operations sought to encourage Jill Stein supporters to make it to the polls, even though she had no chance of winning. On Election Day, Democratic turnout for Hillary Clinton in both states was lower than in previous elections. Russia's pump for Trump, derision of Clinton, advocacy for Sanders, and slight nudging for Stein could easily account for this small turnout differential.

"But I'm not on social media, so I wasn't influenced by the Russians" consistently rates as a top counter to ideas of Kremlin shaping of American minds. These voters mistakenly believe that the social media and mainstream media worlds don't cross, but major news networks increasingly rely on social media to generate their stories. After the election, one set of researchers at the University of Indiana undertook the due diligence of analyzing the relationship between social bots, fake news, mainstream media, and influence.[7] After analyzing fourteen million messages spreading four thousand claims on Twitter from before and after the 2016 election, the Indiana team discovered that social bots provide an essential amplification effect, making false claims go viral across Twitter. These bots also directly targeted influential users on social media, such as political candidates and the media reporters and producers who cover these candidates. Automated social bots influence what mainstream journalists report on.

Whether a voter uses Twitter or not, many of the stories one consumes in print, radio, and television media originate on social media. Not using Twitter does not prevent one from being exposed to what goes viral on Twitter. Any casual television watcher or radio listener has been exposed to the endless debates on President Trump's Twitter usage since assuming office. Even today, the president's Twitter dialogue may be the largest topic of discussion on many mainstream outlets. No border separates social and mainstream media—the two are symbiotic and synergistic.

Social media companies came under fire from Congress and the American public in 2017 for not detecting the Kremlin's manipulative measures on their platforms. Facebook seemed to be the first to catch on to the shenanigans coming from the St. Petersburg–based Internet Research Agency. In response, the platform shuttered thousands of accounts prior to the 2017 French presidential election. In October 2017 the social media giant revealed that more than a hundred confirmed Russian-backed pages had spent more than $100,000 on ads. Facebook's general counsel, Colin Stretch, described these ads as "deeply disturbing . . . an insidious attempt to drive people apart" using race, religion, gun rights, and gay and transgender issues to inflame social and political divides. These known Russian-backed accounts posted more than 80,000 pieces of content and may have touched up to 126 million users.[8] Fake Russian social media accounts created 129 Facebook events, which reached 300,000 users and gained 62,500 users saying they'd attend and another 25,800 users expressing interest in the events. One event page, called Heart of Texas, called for the secession of Texas from the United States. It initiated a "Stop Islamization of Texas" event in Houston, trying to rally protesters at an Islamic center on May 21, 2016. Then the trolls hosted a United Muslims of America page to draw people to another Houston event called "Save Islamic Knowledge," occurring at the same time. The Kremlin used Facebook to pit competing factions against each other at a time and place of its choosing, and it attempted this among nearly all American divides around the country.[9]

Google found Russian influence at a smaller level than Facebook, but the manipulation was equally disturbing. Internet Research Agency accounts bought $4,700 in advertising and, through eighteen channels, hosted more than 1,000 videos receiving more than 300,000 views.[10] The depths of the Kremlin's efforts were discovered when it was revealed that it had created a YouTube page called Williams and Kalvin. The page's videos showcase two black video bloggers, with African accents, appearing to read a script stating that Barack Obama created police brutality and calling Hillary Clinton an "old racist bitch." Some dismissed the effort as a phony failure, but the Williams and Kalvin Facebook page garnered 48,000 fans.[11] Russian influence operators employed most every platform—Instagram, Tumblr, even Pokémon Go—but it was the Kremlin's manipulation via Twitter that proved the most troubling.

Twitter's revelations of Russian meddling have been slow and concerning, and they continue to grow with each month. As Facebook and Google returned their findings to the Senate Select Intelligence Committee, Twitter seemed unaware of any Russian influence connections on its platform. In early October 2016, Twitter initially knew of a couple hundred accounts connected to the Kremlin. That number had grown by ten times by the end of that month.[12] On January 31, 2018, Twitter upped the count to 3,814 Internet Research Agency–backed accounts, along with 50,258 automated accounts that were "Russia-linked and Tweeting election-related content during the election period." I estimate this to be only a fraction of the Kremlin's accounts. Twitter has since notified more than 1.4 million users it believes might have interacted with Russian government-linked accounts. John Cornyn of Texas, the second-highest-ranking Senate Republican, received one of the notifications.

Thousands of Russian-backed Twitter accounts influenced the

presidential election cycle, but an example of how a single account can influence debate comes from @TEN_GOP. This Russian-operated account posed as a Tennessee Republican Party account and gathered 136,000 followers, many multiples greater than the account of the actual Tennessee Republican Party, which tried to have the impostor account closed. This single Russia influence account received retweets from Brad Parscale, the Trump campaign's digital director, Donald Trump Jr., who repeated its bogus voter fraud claims, Michael Flynn, Michael Flynn Jr. (more than thirty times), and conservative pundit Ann Coulter (fifteen times).[13] The Kremlin's @TEN_GOP account received citations from nearly every conservative outlet heading into the election. Kevin Collier at *BuzzFeed* noted, "Fox News quoted an @TEN_GOP tweet in at least three stories . . . The Daily Caller itself quoted it in six stories, . . . Breitbart mentioned it in seven; Infowars in four; RedState in eight. . . . The Gateway Pundit . . . cited the Russian account in 19 different stories."[14] I'll say it again: this was just one of thousands of accounts operated by Russia in the lead-up to the election pushing for Trump and hating on Clinton.

But even so, did Trump directly collude with the Kremlin? President Trump's opponents have advanced the conspiracy that the commander in chief worked with the Russians to gain the nation's top spot and that he now does Putin's bidding. Tales of Kremlin operatives compromising Trump family members, perpetuated by the Christopher Steele dossier and left-wing conspiracy theorists, have littered the internet. Investigative journalists continue to trace the connections between Trump Tower and Moscow. Trump's own behavior—his public support for Russian foreign policy positions, his adoration of Putin and failure to articulate a Russia policy to counter the Kremlin, despite its well-documented meddling, and the abrupt firing of FBI director Comey, among other bizarre behaviors—strengthens claims of collusion.

Trump's supporters offer several counterarguments. One says that Trump naturally allies with the Kremlin; their views and policy positions converge. A separate argument posits that Trump is a political opportunist, his behavior consistent with his deal-making strategy in the business world. A more recent argument comes directly from the White House. Jared Kushner appeared behind closed doors with the Senate Select Intelligence Committee and released a public statement discussing all of his contacts with Russian diplomats, bankers, and envoys during the campaign. The statement, in sum, points to inexperience, disorganization, and chaos—in essence, it blames the Trump campaign's incompetence for the way it stumbled into Russian influence.

The debate over whether the Trump team colluded with Russia ultimately is a debate about the word *collusion*—secret or illegal cooperation or conspiracy that seeks to cheat or deceive others. Collusion comes in degrees, however, and ultimately the Kremlin would prefer not to engage in direct collusion. Enlisting the Trump team as direct agents infiltrating the highest ranks of government would be an immediate provocation for war between the United States and Russia.

When Vladimir Putin sat down at a table during an RT dinner in Moscow in December 2015, he enjoyed the company of two Americans: Michael Flynn, who would later join the Trump campaign, and Jill Stein, a presidential candidate for the Green Party,[15] a subset of liberal America with a penchant for U.S. government conspiracies. In Flynn, Russia saw an ally in the fight against terrorism and a man amenable to taking financial payment. Flynn compromised himself by appearing at that dinner at all, providing the Kremlin with information it could use against the retired general should he go against its wishes. Flynn later appeared at meetings in the United States with Ambassador Sergey Kislyak and, in his zealous and aggressive pre-election plotting, made a phone call to the ambassador in advance of President Obama's sanctions against the Russian government. Flynn's obfuscation about this monitored phone call led to his firing only weeks into his new job as national security adviser. The Russians didn't direct Flynn; Flynn did their bidding for them.

Trump's inexperience also led him to hire Paul Manafort as his campaign manager. An experienced political hand, Manafort had spent the better part of the previous decade consulting and operating in Ukraine, at times for now-exiled president Viktor Yanukovych, a well-documented, cultivated ally of Russia. Manafort led the planning and operations of the Republican National Convention, where again Ambassador Sergey Kislyak appeared, encountering future attorney general Jeff Sessions. During the same convention, Trump staffers stripped language from the party platform that supported the provision of defensive weapons to Ukraine, directly countering the notoriously hawkish Republican stance against Russia.[16] Manafort, an unpaid campaign manager for Donald Trump, departed the campaign shortly after repeating a Kremlin conspiracy on CNN and being tied to a ledger payment from Ukraine's pro-Russian Party of Regions in the amount of $1.2 million.[17] As of this writing, reports from CNN assert that U.S. intelligence agencies intercepted Russian operative communications discussing efforts to work with Manafort and provide information damaging to Hillary Clinton.[18] Manafort has been indicted by the special counsel's investigation and awaits trial on a range of money-laundering charges. Even if Russian connections to Manafort prove to be true, no evidence suggests that Russia colluded directly with Trump as of this publication.

\\\\\\\\\\\\\\\\\\\\\\\\\\\\

Perestroika and glasnost—economic liberalization and opening up of information—crumbled the Soviet Union, but two decades later they opened the way for Russian active measures far more successful than those of their forefathers. The closed Soviet economy provided no method for openly and legally incentivizing accomplices. KGB agents instead had to recruit operatives and issue payments in the conduct of active measures. Today, Russia's active measures economically influence America. The Kremlin doesn't need to pay the Trump team and its envoys when mutually beneficial business arrangements have naturally brought the two camps together. President Trump's son, Eric Trump, smugly noted that a significant portion of the investments they received were Russian. Donald Trump attempted for years to invest in Russia, even hosting the Miss Universe pageant in Moscow in 2013—an event now shrouded in conspiracy. Even as he began his run for president in 2015, Trump's businesses were seeking to build a tower in Moscow.[19]

Carter Page is the perfect embodiment of Kremlin and Trump economic convergence. Page, an oil-and-gas expert and investor, stumbled into Trump Tower during the campaign and became one of the nominee's foreign policy advisers. But in the summer of 2016, Page stepped onto a Moscow stage and gave a lecture that sharply criticized the United States and promoted Russian economic and foreign policy positions. Soon after, the Trump team shed Page and distanced itself from him. As with Manafort and Flynn, Kremlin influence connections to Page have surfaced that date back to 2013. An espionage case from New York City in 2013 convicted two Russian spies. Those proceedings illuminated Russia's intentions to recruit an American businessman amenable to Kremlin policies. The court proceedings anonymized the name, but journalists believe this individual was Carter Page.[20] Page claims that he wasn't aware of this recruitment, and he's likely telling the truth. From Russia's perspective, why coerce an asset if the target willingly pursues the Kremlin's interests?

The most egregious and foolish connection between the Trump campaign and the Kremlin occurred on June 9, 2016, in New York City.[21] Rob Goldstone, an English music producer and friend of the Trump family, emailed Donald Trump Jr. about setting up a meeting at Trump Tower with Natalia Veselnitskaya, a Russian lawyer, and Rinat Akhmetshin, a Russian immigrant with ties to a range of Kremlin intelligence. "Russia—Clinton—private and confidential," the email's subject line read.[22] Donald Trump Jr. took the meeting under the auspices of receiving damaging *kompromat* on Clinton, but the meeting devolved into a lobbying effort to repeal the Magnitsky Act, a congressional law passed to prevent those responsible for the 2009 death of whistleblower Sergei Magnitsky from gaining access to the United States and utilizing the U.S. banking system. Trump Jr. states that nothing of consequence was discussed. Veselnitskaya claims she was summoned to Trump Tower. Again, Russia created an incident where it could influence the narrative, sow discord in America, and suggest incompetence.

Russia can influence its intended targets through emissaries, envoys, and oligarchs. Should a target shun Russian advances or even rebuff the Kremlin's overtures in private, a public meeting with a target provides a real world encounter Russian propagandists can reframe to discredit an opponent through conflicting stories or allegations. For example, in the case of Donald Trump Jr., Natalia Veselnitskaya claimed during an NBC interview that it was the Trump team, not her, who sought damaging information on Hillary Clinton. Jared Kushner suffered similar challenges after accepting a pre-inauguration meeting with Ambassador Kislyak. The media later reported that Kushner sought to establish secret back-channel communications from the Trump team. In both cases, accepting these Russian meetings resulted in the media running with a narrative that was damaging to the Trump presidency.

As with all of Russia's active measures, the essential element is plausible deniability: the actions and outcomes of Russian influence should appear as natural occurrences. Guccifer 2.0, the slipping of hacked DNC and Podesta emails to WikiLeaks, a Trump Tower meeting regarding the Magnitsky Act, multiple chance encounters with Ambassador Kislyak, an RT dinner with a retired U.S. general and intelligence chief—each can be explained away. Putin, when asked about Russian meddling in the U.S. election, said the Russian government had nothing to do with such interference but that it's possible that "patriotic hackers" in Russia may have undertaken actions independently.[23] By muddying the waters, Putin indirectly admitted that some Russians may have been involved but also offered plausible deniability of his state's direct involvement.

Ensuring plausible deniability requires a sustained campaign far beyond any single active measures action or achievement. The Kremlin counters attribution claims using three techniques: alternative perceptions, parsing and refuting facts, and counteraccusations. It implements this methodology through a full-spectrum approach employing diplomats, selective information, and proxies to challenge attribution claims. The best example of this approach is the story of Seth Rich, a DNC staffer murdered on July 10, 2016, who became a tragic conspiracy scapegoat among Russia meddling investigations. D.C. police believe that Rich's murder was the result of a robbery gone wrong. WikiLeaks' Julian Assange, however, issued a reward for details regarding the murder, suggesting that he may have been the source of the DNC email leak. Alt-right websites and conspiracy theorists on 4chan and Reddit posited the same explanation: Rich knew of DNC corruption, they said, and was a closet Bernie Sanders supporter seeking to air the truth.[24] RT and Sputnik continued stirring up these allegations through public news stories, amplifying WikiLeaks' conspiracies,[25] and Russian diplomats joined in this counternarrative as well. On May 19, 2017, the Russian embassy in London tweeted, "#WikiLeaks informer Seth Rich murdered in US but 🇬🇧 MSM [mainstream media] was so busy accusing Russian hackers to take notice."[26] The Russian troll army amplified this conspiracy even further, retweeting claims of a U.S. government cover-up. A Fox News story repeating the claim that Rich may have been the source of DNC leaks has since been retracted, and the news outlet has come under suit by a private detective wrongly cited in an article. A year after Seth Rich's murder, no evidence has emerged to support the conspiracy about his being the source of the DNC email leak, but the challenge to U.S. government explanations of DNC hacking lives on.[27]

Russia's vast cybercrime underworld adds to the Kremlin's plausible deniability and provides Putin's minions with a strategic edge against his adversaries. The National Security Agency (NSA) and, more recently, the new U.S. Cyber Command (CYBERCOM) struggle mightily to recruit, train, and retain hacking specialists in their ranks. Vast differences in pay and culture make government work unpalatable for the highly skilled hackers of America. Even if interested in serving their country, most top potential recruits to America's cyber ranks have either committed a crime by breaching networks and databases or used a controlled substance, which bars them from government employment. Both of these stipulations routinely disqualify patriotic Americans willing to serve the country's intelligence services.

Russia and its Eastern European neighbors maintain highly educated citizenry who—when finally exposed to the world of information technology after the end of the Soviet Union—took to computer science in droves. Computer application coding became a burgeoning industry after the fall of the Iron Curtain, and Western countries outsourced their technological needs to the former Soviet republics, valuing their skill and low prices. At the same time, Putin's Russia, struggling economically and moving toward kleptocracy, allowed organized crime to run wild, and these groups drifted toward a nearly infinite new enterprise of cybercrime and scooped up a rising pool of talented young programmers.

By the end of the first decade of the twenty-first century, a large portion of global cybercrime came from Russia. The Kremlin largely turned a blind eye to the cybercriminals within its borders, rarely arresting international perpetrators unless they provided a bargaining chip in other state-to-state negotiations. Putin's cronies and criminals instead implemented a selective black-market tax on the cyber underworld, taking bribes from Russian syndicates, which allows them to continue their hacking as long as they avoid targeting domestically and inflict no damage on the Russian regime. Russian intelligence services today use this cybercriminal underworld as a means of hiding their activities.

Sometimes the Russian military advertises jobs for coders in the same way that U.S. military contractors seek out new employees: by posting job ads on social media and hosting recruiting sessions at colleges. Mandatory conscription provides a separate vector for the Russian military: those wanting to avoid the tougher challenges of standard military units could join a "science squadron," where they might employ existing cyber skills or learn new ones. But the best in Russia, much as in America, have found greater incentive to stay outside the Kremlin's grasp—profits and notoriety abound in the Wild West of Russia's cyber underworld. The Russian military and intelligence agencies further bolster their ranks through coercion and contracting.

When the Obama administration announced Russian sanctions at the end of December 2016, a curious mix of criminal hackers and cyber companies appeared on the list, alongside known Kremlin intelligence outfits previously connected to election meddling. Evgeniy M. Bogachev is the most wanted cybercriminal in the world. He controls the GameOver ZeuS botnet, which has drained millions upon millions of dollars from bank accounts around the globe. Conveniently, the same botnet provides access to every infected computer, and in March 2014 Bogachev's botnet began searching for politically sensitive information, helping look for intelligence in support of Russia's invasion of Crimea. It's believed that sometime during this era, Bogachev received free passage to conduct cybercrime around the world, in exchange for Russia's internal security service, the FSB, being allowed to lean on his services and his botnet to conduct espionage when needed. Today, Bogachev lives openly in Anapa, Russia, sailing on the Black Sea, and Russian authorities claim no grounds for arresting him, since he's never robbed a Russian citizen.[28]

Alisa Shevchenko, a former employee of Kaspersky Lab, founded the company Zor Security (formerly known as Esage Lab), a business that she openly advertised as an offensive security research firm focused on finding vulnerabilities rather than fixing them. She has even received credit from the U.S. government for identifying critical weaknesses in code. Shevchenko and Zor Security's connections to election meddling remain unclear, but an examination of YouTube videos places her as a founding member of a white-hat hacker collective known as Neuron Hackspace, in Moscow. A fellow cofounder is none other than Dmytro Oleksiuk, once a technician for Esage Lab in 2012, who authored the base code for a multipurpose hacking Trojan known as BlackEnergy—computer coding later used to black out western Ukraine just before Christmas 2015. Dmytro claims no knowledge of or participation in the later employment of his base code, and Alisa claims no connection to the Russian government, despite curious contracting relationships with the Kremlin.[29]

The Kremlin's chef, Evgeniy Prigozhin, found his catering company on a separate sanction list fought by President Trump in the summer of 2017. Concord Catering supplies Moscow's public schools, but Prigozhin allegedly also provides the funding for the Internet Research Agency, Russia's social media troll farm in St. Petersburg. The summer 2017 sanctions list also named two administrators, two organizations, and the Moscow bike center tied to Kremlin-funded motorcycle gang the Night Wolves, who play a prominent role in the theatrics of Russian active measures.[30] The Night Wolves act as a visible proxy for Putin, who sometimes appears at their rallies. The biker gang stormed into Ukraine as the Russian annexation of Crimea began. Flying Russian flags, they fought alongside Russian separatists, and they act at times as a civil affairs arm for winning over local support among the insurgency. Their antics inspire alt-right audiences and advance anti-Western sentiments.[31] Bogachev's botnet, Esage Lab contracting, Prigozhin's troll army, and a ground force called the Night Wolves—criminals, hackers, harassers, and bikers—combine to create the Kremlin's active measures proxy army, doing Putin's business with a veil of plausible deniability.

\\\\\\\\\\\\\\\\\\\\\\\\\\\\\\\

The height of intrigue over Trump's possible collusion with or compromise by the Russians came in January 2017, when ex-MI6 officer Christopher Steele's Trump dossier was published by *BuzzFeed*. The dossier contained what Trump haters hoped for: juicy tales of sexual compromise mixed with Kremlin deliberations over how far to back Trump over Clinton. Some portions were verifiably true, others were likely false, and many parts remain debatable. The best interpretations of the Steele dossier come from fellow veterans of the British Secret Intelligence Service, the writers John le Carré and Ben Macintyre.[32]

When asked, "Do you think the Russians really have something on Trump?" Ben Macintyre provided the best explanation of what Putin is really up to with his election meddling:

> Yes, *kompromat* was done on him. Of course, *kompromat* is done on everyone. So they end up, the theory goes, with this compromising bit of material and then they begin to release parts of it. They set up an ex-MI6 guy, Chris Steele, who is a patsy, effectively, and they feed him some stuff that's true, and some stuff that isn't true, and some stuff that is demonstrably wrong. Which means that Trump can then stand up and deny it, while knowing that the essence of it is true. And then he has a stone in his shoe for the rest of his administration.
>
> It's important to remember that Putin is a K.G.B.-trained officer, and he thinks in the traditional K.G.B. way.

Le Carré then added:

"As far as Trump, I would suspect they have it, because they've denied it. If they have it and they've set Trump up, they'd say, 'Oh no, we haven't got anything.' But to Trump they're saying, 'Aren't we being kind to you?'"

Regardless of whether Trump won because of Russia or through his disruptive style and populist message, Russia achieved a major victory when Trump became president.

Never in American history has a U.S. leader taken such a proactive and warm approach to Russia or its predecessor, the Soviet Union. Public support for Russia prior to Trump's ascendance traditionally arose from the American left wing, those more amenable to socialist and Communist narratives traditionally and recently those mobilized by the Edward Snowden disclosures. But with the rise of Trump, a rapid and sizable shift of support for Putin and company has come from the American right wing—the same GOP that led the way against Communism and claimed victory over the Soviet Union under Reagan and Bush. In a February 2017 Gallup poll, Putin's favorability rating showed a 166 percent increase among Republicans and a 92 percent increase among independents just since 2015. Among all Americans, Putin's favorability doubled in just two years. Nearly a quarter of all Americans now approve of a foreign leader who just launched the greatest assault on democracy in history.

For those surprised by this American affection for authoritarianism, don't be. Political science professor Seva Gunitsky recently reminded Americans that fascism and Hitler were equally popular during the 1930s. Gunitsky compared U.S. public opinion in the 1930s with opinions today, noting that, "as late as July 1942, a Gallup poll showed that 1 in 6 Americans thought Hitler was 'doing the right thing' to the Jews. A poll found that nearly a fifth of Americans saw Jews as a national 'menace'—more than any other group, including Germans."[33] In both the 1930s and today, the swing in American support against Hitler and for Putin likely arises from the president more than any other factor. Roosevelt declared war on Hitler, and Trump's rise came with a public embrace of Putin. Americans like to believe they are independent thinkers, but rapid public opinion shifts suggest otherwise.

Even beyond simple favorability toward a dictator who may have accumulated more wealth than any other man in the world, Russia's influence efforts and rising social media use track closely with declining American faith in democracy. America's millennial generation, when asked if it's essential to live in a democracy, agreed only 30 percent of the time, a 45 percent drop from those born fifty years before them.[34] In a separate 2014 poll, support for democratic political systems declined by 18 percent between 2006 and 2014.[35]

Trump supporters, who more often encounter Russian propaganda than the broader population, also theoretically supported the rollback of amendments limiting presidential terms. When asked, "If Donald Trump were to say that the 2020 presidential election should be postponed until the country can make sure that only eligible American citizens can vote, would you support or oppose postponing the election?" a majority of the 650 Trump-supporting Republicans surveyed said yes. Almost half thought Trump had won the popular vote, and more than two-thirds believed that millions of illegal immigrants had voted and that voter fraud happens regularly.[36] These baseless falsehoods came from Trump's mouth and were promoted by the Kremlin. These surveys remain hypotheticals, and it's unlikely that such constitutional changes or postponement of voting would stand up to checks by the legislative and judicial branches. But at no time in recent U.S. history have such authoritarian principles—suspension of democratic norms and suspension of constitutional provisions—been considered or even mentioned in political discussions. All of these changes, allegations, and considerations eerily mirror the actions of a dictator Trump openly admires: Vladimir Putin.

Upward trends for shying away from democratic governance may not adequately demonstrate the magnitude of Russia's influence success, but two anecdotes since the election provide a stark reminder of just how much things have changed in a very short time. On May 13,

2017, a startling sight occurred in Charlottesville, Virginia.[37] A mob of white nationalist men toting torches descended on a statue of Robert E. Lee, the Confederate Civil War general, protesting the planned removal of the artifact. The crowd chanted and shouted, "White lives matter" and "No more brother wars," strangely adding "Russia is our friend" to the list of slogans.[38]

Imagining a time during the Cold War when white men might have gathered to chant, "The Soviet Union is our friend" seems impossible, but in the post-Trump world, Virginia has become a common site for pro-Putin rhetoric—a conservative state home to many current and retired servicemen who fought the Soviets during the Cold War. Three months later, white nationalists returned to Charlottesville for another torch-wielding rally, and this time violence broke out between white nationalist protesters gathering at the same statue and counterprotesters seeking to challenge their calls. Bickering boiled over into tragedy when a young man from Ohio, a white nationalist, barreled his car into a crowd of counterprotesters, killing a woman and injuring many more. Alongside the violence, protesting, and social media coverage, another strange video surfaced. White nationalist American men interviewed during the protest boasted of their support for Syrian president Bashar al-Assad. Footage posted by Brandon Wall, using the Twitter handle @Walldo,[39] showed a spirited white nationalist clad in a "Bashar's Barrel Delivery Co." T-shirt and yelling into the camera, "Fight against the globalists. Assad did nothing wrong . . . Two chemical bombs would have solved this whole ISIS problem."[40] In roughly fifteen seconds, two of the Kremlin's most highly promoted themes were repeated and displayed in the middle of the United States.

Putin's greatest achievement in these social-media-powered active measures will ultimately be his ability to unite previously disparate alt-right communities through persistent information campaigns. The Kremlin won audiences in the lead-ups to elections and has maintained them long beyond the close of polls. In the UK's Brexit vote, the U.S. presidential election, and the French and German contests of 2017, Russia has effectively leveraged online white nationalist communities as indigenous counters to their American and European political rivals.

Across all of these alt-right audiences, anti-immigration programming has remained a consistent feature of RT and Sputnik news stories. Unrelenting refugee flows into Europe are an enduring theme echoed in Kremlin media. Russia has amplified allegations of refugee crime and its drain on European governments. Anti-immigration and anti-refugee sympathies disseminated by Kremlin media outfits further support Russia's broader calls for nationalism over globalism, and for breaking the European Union and the NATO military alliance into pieces, which would provide Putin space to maneuver successfully against any other country in a one-on-one fashion.

The closer one gets to Moscow, the more aggressive active measures become, with greater combinations of physical actions designed to drive cyber influence discussions. Montenegro's entrance into the NATO alliance on June 5, 2017, met significant challenges from Russia.[41] The former Yugoslav republic's coastal ports represent the only stretch of the Mediterranean not under NATO control, and in recent years it has become a hub of Russian investment and tourism.[42] Less than a month before Trump and Clinton squared off in America, the Montenegrin election showcased the Kremlin's reach and the lengths to which it would go to employ active measures to influence strategic electoral outcomes.

On election day, Montenegrin authorities arrested twenty men, mostly ethnic Serbs, dressed as police officers, who planned to break into Montenegro's parliament, kill Prime Minister Milo Djukanovic, and claim that electoral fraud had robbed pro-Russian parties connected to Putin of victory.[43] A month after the detentions, prosecutors in Montenegro alleged that nationalists from Russia had organized the Serbian mini-invasion of their country, with two individuals, Eduard Shirokov and Vladimir Popov, identified as the plot's masterminds, on assignment from Russia's military intelligence arm, the GRU. Among the Serbians, Aleksandar Sindjelic, head of the nationalist Serbian Wolves organization, an outfit similar to the Russian Night Wolves, allegedly received 200,000 euros from the GRU operatives to undertake the coup attempt.[44]

Oddly, during the NATO summit in May 2017, President Trump shoved aside the newly elected prime minister of Montenegro, Dusko Markovic, to get in the front of a pack of dignitaries. Surely Trump's brush with Markovic was just a coincidence—or was it a signal from Putin's newest ally, Trump, who has been a divisive force in NATO unity since inauguration?[45]

Elections and politics remain a constant political vector for Kremlin meddling, but Putin's information war doesn't stop in those years between votes. Russian active measures leading up to an election or following an election focus on social issues, and immigration remains a chief lever for winning over foreign audiences. Lisa F., a thirteen-year-old German-Russian girl with dual citizenship, went missing on her way to school in Berlin on January 11, 2016. A little over a day later, the girl resurfaced, claiming she had been abducted by three men of "southern" or "Arab" origin, none of whom spoke German very well.

Russia's state-sponsored media quickly ran wild, broadcasting numerous and repeated stories of immigrants raping Lisa F.—asylum seekers in Germany who'd held her as a sex slave. Germany's Russian community, numbering nearly two million, reacted strongly to the rape story, and within a week Russian émigrés were protesting in several German cities and at the office of German chancellor Angela Merkel, sporting banners that read, "Our children are in danger" and "Hands off my child."[46] The Russian foreign minister, Sergey Lavrov, blamed the German government for squashing the story out of political correctness.[47]

The case of Lisa F. quickly fell apart as investigators sought out the perpetrators. German authorities pulled her cell phone records and traced her communications to an older teenager with whom she'd previously had relations. Within hours of her initial claims, Lisa F. gave a slightly different accounting of her whereabouts, saying that she'd voluntarily gone with the men. Months later, investigators determined that she'd stayed the night with a friend and fabricated the entire abduction-and-rape story.[48] Lisa F.'s family remained steadfast in promoting the girl's original story, though, and likely many Russians in Germany still believe the first account they heard, not knowing that later the entire conspiracy proved nothing more than a fabrication.

A Russian GRU textbook on psychological warfare dating from 1999 recently surfaced, explaining active measures techniques in great detail. A student of this course, known as Dmitry, recalled to the *Moscow Times* "the specific black propaganda techniques taught in his textbook." He described one ploy used: "During the Chechen conflict, Russian psychological warfare experts spread rumors that foreign fighters had raped the 13-year-old daughter of a Chechen village elder. The rumors helped sow discord between Chechen fighters and Arab Islamist volunteers, undermining the unity of the rebels."[49] Whether it's Berlin, Chechnya, or America, the playbook is old; the medium is new. Social media provides a sharp, devastating new edge.

At this point, one might conclude that Russia's social active measures have been diabolically successful. This isn't necessarily the case, particularly since the U.S. election. With each active measures action, Russia's hand and the game it plays are exposed a little more. From the start, Russia's decision to launch cyber-enabled active measures to influence elections has been a campaign rather than a single battle—Ukraine in 2014, Brexit in 2016, the U.S. presidential election of 2016, France's presidential contest of 2017, and finally the German federal election of 2017. In these latter two contests, France and Germany have proven a bit more resilient.

Just before France's two-day media blackout leading up to the presidential election, Jack Posobiec, the Washington bureau chief of the Canadian conservative outlet *The Rebel*, tweeted, "Massive doc dump," pointing everyone to hacked emails from French presidential candidate Emmanuel Macron.[50] For months, Russian hackers and state-sponsored media employed their well-run cyber active measures against Macron, in favor of their preferred candidate, Marine Le Pen—a fellow traveler of Putin, openly appearing alongside him in photos and going so far as to visit the Kremlin only two months prior to the election.

But the French electoral system structurally inoculates itself against Russian meddling. The first round of voting, on April 23, 2017, saw four candidates each win around 20 percent of the vote. For the Kremlin's influence system to push Le Pen to the top, it needed to push aside three challengers. Even after the field narrowed to Macron and Le Pen, the final vote went to the polls two weeks later, rather than ten months later, as in the United States. This provided Putin's meddlers with only a narrow window in which to release compromising information for the mainstream media to discuss and social media to circulate.

There was no time for the release of the Macron emails on May 5, less than two days before the final vote, to gain any traction in people's opinions, particularly due to the media blackout. Instead the Macron emails sought to undermine his mandate to govern post-election, similar to the allegations of voter fraud in the U.S. context.

France, in general, boasts a society less receptive to compromising data dumps. While sexual indiscretion has been a principal angle for smearing opponents in the United States, France's last two presidents, Nicolas Sarkozy and François Hollande, both were thought to have had affairs with other women. Compromising information released by Kremlin hackers would have to be egregious and criminal in order to offend French sensibilities and shift opinions. The troll army certainly sought to diminish Macron's chances, but its reach with overt RT and Sputnik news outlets remained limited, and the social media influence in French didn't appear as robust or as effective as the English-language campaign employed on Americans. Even if it had been, social media news consumption in France (and in Germany) measures only about half of that in America. Europeans, for now, especially older voters, get their news the old-fashioned way: through traditional news outlets and conversations with friends and family, trusted sources of known origin and stronger track records.

Putin's greatest prize may be Germany—the remaining dominant player in the European Union and NATO. Putin failed to help replace Chancellor Merkel and install a sympathetic supporter, but he did help increase the share of parliament members from the right-wing, Russia-sympathetic Alternative for Germany (AfD) party. Germany's domestic intelligence agency claimed that Russia's GRU hackers, the same APT28 credited with hacking the DNC, broke into Germany's lower house of parliament, the Bundestag, in May 2015, along with other offices of Merkel's Christian Democratic Union. Germany's much larger Russian immigrant population carries stronger sympathies for its homeland than France's smaller émigré community, as indicated by its large showing at the Lisa F. protests. Germany's alt-right audiences' draw to Russian propaganda has proven far stronger than its counterparts' reaction in France, and the Kremlin's social media troll army appears more skilled in both understanding the German audience and communicating in German.

On March 30, 2017, I testified to the Senate Select Intelligence Committee regarding Russian influence operations leading up to the election. Since then, I've briefed nearly every agency of the U.S. government in one form or another. When I started the Russia research, I expected it to go the way of my deep dives on al-Qaeda and the Islamic State: a research paper would lead to a project examining Russian active measures. A think tank or a government agency would offer some funding to map out the landscape. Al-Qaeda or ISIS tracking and analysis studies came fairly easily. Everyone wanted a Russia brief, so I thought the process would mirror my counterterrorism work. It turns out that tracking Russia's social media active measures has been the best research I've ever been a part of, and also the worst business decision I've ever made.

Immediately following Trump's election, I sketched out my recommendations for how to counter Russia's active measures campaign. Building on lessons learned from the Cold War, I felt that a modern update to the techniques utilized by the U.S. Information Agency could be employed on social media—that is, if the U.S. government wanted to do anything about it. The first step was to raise awareness of Russian active measures by outing the propaganda outlets disseminating Kremlin themes. But since the end of the Cold War, no U.S. agency has assumed responsibility for this key role, and no agency I pitched the concept to seemed interested in pursuing the mission—all seemed paralyzed by Trump's inaction on Russia throughout 2017.

During the war on terror, I worked "by, with, and through" the U.S. government. To fight Russia from my office and a laptop, I would need to go "over, under, and around" the U.S. government. Westerners wanting to protect their democracies need to bypass their governments and create their own online counterinsurgency, and here in civil society we may all develop the best solutions. The first step: show the world what the Kremlin is peddling. Arm the public with information and awareness, challenge Russian conspiracies and narratives, and have them challenge Russia's bots, propagandists, and authoritarian supporters.

My colleagues and I initiated a new approach, and this time I wasn't going to be alone messing with an enemy on social media. We'd been watching feeds for years that tipped us off to the Kremlin's disinformation spreading on Twitter. If the public could see what we were seeing, it could avoid consuming or inadvertently promoting Russian propaganda. Journalists could research the outlets pumping Kremlin themes, and citizens could work to discredit false personas pushing fake news. J. M., Andrew, and I teamed up with Jonathon

Morgan to create a real-time dashboard displaying the summation of the key Twitter accounts we monitor. We initially planned to go it alone, but the German Marshall Fund of the United States' new Alliance for Securing Democracy, a nonpartisan effort to counter Russian influence worldwide, emerged to back our pilot project. The dashboard showcased two components of Russian social media influence: the stories the Kremlin's overt English broadcasting media outlets were spreading, and the topics and stories Russia's Twitter troll networks—a mix of overt Russia supporters and social bots—promoted to American audiences. We named the platform Hamilton 68, harking back to the *Federalist Papers*, where Alexander Hamilton, writing under the alias "Publius," warned of foreign meddling in American elections.

"These most deadly adversaries of republican government," Hamilton wrote, "might naturally have been expected to make their approaches from more than on quarter, but chiefly from the desire in foreign powers to gain an improper ascendant in our councils . . . by raising a creature of their own to the chief magistracy of the Union."[51]

Hamilton foresaw the vulnerabilities of a free society—how an outside meddler could harm Americans—long before social media arrived to open up the Kremlin playbook in America.

The Hamilton 68 dashboard launched on Wednesday, August 3, 2017. In only a few days, the website garnered hundreds of thousands of visits, and we aggressively tried to explain how the page worked, what one can learn and observe. Meanwhile, I waited for a Kremlin counterattack, a public data dump or smear campaign. By and large, the feedback proved positive, and the public took to studying the top and trending Twitter topics pushed by the Kremlin. Of course, some on the left accused us of McCarthyism and questioned our methodology, suggesting that we somehow were trying to malign unnamed Twitter accounts. Right-wing trolls, meanwhile, shouted a well-worn label. "You are part of the deep state," they would tweet.

We knew we had been successful when five Sputnik English articles discrediting the Hamilton 68 dashboard and the German Marshall Fund appeared in the first forty-eight hours after launch. Each article strategically sought a new angle to undermine the dashboard along one of two themes: the dashboard was either laughably incompetent or extremely weak in its capability. "Social Media Activists Mock Project Tracking Sputnik, RT Accounts" cited overt Twitter supporters of the Kremlin. "US-Funded Service Tracking Sputnik, RT Attempt to 'Monopolize Truth'" quoted an unknown French lawmaker. But the best response was "Ex-Russian Ambassador to US Kislyak Surprised by US Efforts to Track Sputnik, RT." Kislyak, the diplomat whose hand tainted so many in Team Trump, said "he had not expected that the level of the western countries' capabilities to articulate threats would be at such a low level."[52] As Sputnik pushed out a total of eight posts maligning the Hamilton 68 dashboard, I watched as the links to those Sputnik articles climbed into the "Trending URLs" tracked on the Hamilton 68 dashboard. I felt even more confident we were on target. The goal of influence is to create a behavior change in one's adversary. Eight articles from Sputnik in less than a month—I think the Kremlin got our message.

Some of our followers visited the dashboard each day, examining the top hashtags and earmarking trending topics. Those with technical skills used Hamilton 68 to employ machine learning to detect and out social bots spreading divisive messages in the United States. One enterprising American began identifying bots tied to Kremlin-promoted hashtags and then repeatedly tweeted anti-Kremlin links at the accounts. Others outed false personas, which would disappear within a few hours or a day. The efforts were small but effective in signaling a turn in the propaganda wars against the Kremlin. A month after the English-speaking Hamilton 68 dashboard launched, we used the same template to offer a parallel dashboard showing Russian influence of the German population leading up to the September 17, 2017, election. The platform used a similar moniker, Artikel 38.

After a year of publicly speaking and writing about Russian influence efforts, I've found communicating the complexity and depth of the Kremlin's deceptive operations difficult to communicate to the public. Confusion over Russian hacking of voting machines and the manipulation of social media by the Internet Research Agency has continued as Congressional committee investigations linger on, selectively leaking information at politically opportune times. The Hamilton 68 dashboard sought to resolve some of this confusion, but explaining its outputs to the public and media outlets remains a challenge and I still look for better ways to articulate Russian subversion to audiences that may not realize the effect its having on them.

U.S. government resources are needed to fund a truly effective effort. Intelligence agencies, Homeland Security, and the State Department would need to rally and coordinate. Instead, on the day before we launched Hamilton 68, *Politico* reported that President Trump's secretary of state, Rex Tillerson, had yet to tap into $80 million in counterpropaganda resources, much of which Congress had set aside for challenging Russian active measures. We hoped to provide a spark, but we still are waiting to see who will carry it forward. Our effort, at the time of publication, remains underfunded, and a side project from our day jobs. It's a start, but not at all what America needs to do against Russian influence. Kislyak was right, and Putin must still wonder, "Why hasn't America punched back?"

Staring at the Men Who
Stare at Goats

"What do I do with this?" Colonel Smith* said to a crowd gathered around a conference table. In his hands was a copy of the *Militant Ideology Atlas*, West Point's Combating Terrorism Center's first major project. Dr. William McCants, fresh out of Princeton University's doctoral program on Middle Eastern history, had just finished a yearlong study with a great team of Arabic-language researchers. Al-Qaeda's online library at tawhed.ws provided nearly every ideological text powering jihad's preachers. Using citation analysis, the West Point team had mapped who referenced whom in all of these militant texts, similar to the way Google's search engine originally tracked web links for its algorithm. Will's team deciphered who the thought leaders behind bin Laden's violent madness were. Thousands of logged citations produced a map of key messages and messengers al-Qaeda terrorists used to enrage their supporters, inspire their followers, and motivate some to join them on the battlefield. The work provided a

---

* A pseudonym.

blueprint, much-needed reconnaissance for launching America's new fight in the "war of ideas" against al-Qaeda.

"Benny, Benny,* what do you think of this? What do we do with it?" Colonel Smith called out for his top deputy, a junior officer whose opinion carried more weight among the staff because he'd read a book on al-Qaeda once and, above all, he had been in this military unit longer than any of the others.

"Sir, I don't know," Benny said, looking disapprovingly at the academics in their suits and ties. "And I'm not sure about putting it on the internet, either." Colonel Smith stared blankly at the table. Before this meeting, the colonel hadn't had a clue about this million-dollar Department of Defense–funded project. Now briefed on it, he didn't know what to do with it, and he didn't want it to bite him in the ass. The academics offered some ideas for what to do with the project's outputs. Chief among them was posting the findings and data on the internet at the Combating Terrorism Center's website. The insights from this analysis could fuel research worldwide and help crowd-source solutions for challenging the underpinnings and justifications of al-Qaeda's new era of terrorism. Experts on militant jihad, world history, languages, and public diplomacy could power their own cost-free, independent research from the project. Colonel Smith concluded the meeting with no conclusion; he'd think about our request for release and likely forget the findings almost immediately.

Weeks of negotiating finally led to the approvals needed to post our research on West Point's website. The military quickly forgot they'd paid for the study—and its results. Academics and researchers visited the work in droves. Within just a few months, the *Militant Ideology Atlas* began surfacing in other studies around the world, fertilizing discussions on what al-Qaeda truly stood for, the group's strengths and weaknesses, and the cracks in its bankrupt ideology. The atlas

---

* A pseudonym.

provided the map, the needed reconnaissance for fighting America's highly touted war of ideas.

I remember this meeting, out of the hundreds I've had since 9/11, only because of where the *Militant Ideology Atlas* landed. Researchers weren't the only visitors to the study; terrorists came, too. Long before I chatted with Omar Hammami on Twitter, al-Qaeda conducted e-counterintelligence on America, perusing the Combating Terrorism Center's website to feed both their narcissism and their fears. At times, the terrorist leaders the center studied, like today's head of al-Qaeda, Ayman al-Zawahiri, publicly complained about the study's reports.

The *Militant Ideology Atlas*, the information weapon we offered to Colonel Smith, the book he gripped and forgot, was recovered by U.S. Navy SEALs on May 2, 2011.[1] Resting on bin Laden's bookshelf were the *Militant Ideology Atlas* and several of the Harmony reports containing bin Laden and al-Qaeda's internal documents. In a note written to a courier, bin Laden issued instructions:

"Please send all that is issued from the Combating Terrorism Center of the American military."

The report is one of the few that actually altered the enemy's thinking, created a behavior change, the prime objective of the war of ideas. Unlike thousands of other reports funded for millions of dollars, it succeeded because it wasn't the work of the D.C. Beltway contracting machine, and because its funders decided to take a risk and make it available to the public. It was a visible victory in the open-source battle with al-Qaeda, and a successful program ultimately killed by the American bureaucracy that once championed it.

<hr />

America sucks at information warfare, absolutely sucks. This isn't necessarily a bad thing. Democracies are marketplaces of ideas. We stand for freedom, liberty, human rights, and peaceful protest, so stopping one thing, like the violent views of terrorists or nefarious

Russian influence of homegrown Americans, gets quite tricky. American values and those of other Western democracies are their greatest strength when shared and promoted—and a major vulnerability in the eyes of those who seek to exploit them. Suppressing ideas undermines American values. And so countering bad ideas, like those that fuel terrorism or authoritarianism, proves vexing, as we tend to believe that the remedy to be applied is more speech, even though we are not entirely sure what to say, how to say it, or who should say it.

Al-Qaeda, the Islamic State, Syria, Russia, or even a political campaign—each knows exactly what it seeks to accomplish on the social media battlefield. The conceptual approach is the first hurdle to successfully countering these bad actors on social media, but the U.S. government doesn't have much consensus about what should be done, how it should be done, and who should do it—the message, the messenger, and the method.

Al-Qaeda's message may have been bankrupt, but it was very clear from the beginning. Western countries, and particularly America, must be attacked, as they are Islam's "far enemy" and they back corrupt Arab dictators (the near enemy) that oppress local Muslims. ISIS, which became the Islamic State, provided equally clear guidance; its name actually spelled out its objective—to create an Islamic caliphate ruled by Sharia law, in line with the time of the Prophet Muhammad. Anyone who fails to follow in this directive or convert to its brand of Islam should be killed. The Russians, in similar fashion, may mask how they influence, but they always know what they seek to achieve with their influence: to reduce American power, creating room for Russia's ascension on the world stage and the pursuit of its foreign policy goals. In Syria, the Kremlin sought to maintain President Bashar al-Assad's rule, secure its own stake in the Levant, restore stability to the country, and quash terrorists. In Ukraine, Russia saw an ethnic Russian population on the strategic Crimean Peninsula that was ripe for a return to Russian rule. Russia's influence campaigns amid Western elections have sought to nudge Western audiences to support the breakup of the European Union and NATO, to promote its own nationalism rather than globalism, and ultimately to undermine democracies.

America's response to each of these influential messages has been difficult to comprehend. When bombs and bloodshed didn't bring an end to al-Qaeda as it migrated to the internet, the U.S. government began pursuing "public diplomacy" to thwart the unending challenge presented by a violent ideology spread across many continents. Conferences were convened and contracts issued for the development of plans and programs to tackle the persistent jihadis plaguing America. By the mid-2000s, the war of ideas had become the new confrontation between the West and al-Qaeda. Billions of dollars soon poured into the creation of counternarratives against all jihadis far and wide.

The Department of Defense started its "soft power" assault on al-Qaeda toward the end of the Bush administration and continued it perpetually through the Obama administration. Defense secretary Robert Gates, who served in both administrations, noted these efforts in 2007, stating:

> The government must improve its skills at public diplomacy and public affairs to better describe the nation's strategy and values to a global audience. We are miserable at communicating to the rest of the world what we are about as a society and a culture, about freedom and democracy, about our policies and our goals. It is just plain embarrassing that al-Qaeda is better at communicating its message on the Internet than America.[2]

Secretary Gates was exactly right, but his words and U.S. government efforts didn't significantly change the problem. Billions in funding didn't result in disenfranchised Muslim men marching toward democracy and away from militancy. Instead, more Western spending on counterpropaganda over the past decade has correlated with the rapid rise of the Islamic State.

American campaigns to promote democracy, civil liberties, and freedoms to disenfranchised Muslim boys landed sideways in countries where religion and tribal loyalties define survival. Repression of civil liberties and freedoms was the staple of the regimes oppressing young Muslim men that trotted off to al-Qaeda—regimes often backed by or partnering with the United States in hard counterterrorism. Inside government, bureaucratic battles over message control and hand-wringing over risk watered down countermessages, creating sanitized flops. Western media contributed to these failures. Newly minted journalists manning blogs, many of them with no understanding of terrorist propaganda, invariably heaped scorn on any government effort. The perceptions of incompetency or ineffectiveness they dropped on civil servants, whether merited or not, would routinely halt any Defense or State Department information campaign.

All the major agencies fighting terrorists launched some sort of information campaign. None were particularly effective, and some even contradicted one another. Today, Russia offers an anti-EU, anti-NATO, nationalist message to sympathetic American audiences. The Trump administration often spouts similar themes. How can America countermessage Russia when President Trump's statements mirror those of President Putin?

America's counterpropaganda doesn't just suffer from a message problem; it also has a messenger problem. Whether it's official spokesmen standing at lecterns or official social media accounts, the West's personas in real life and the virtual world don't resonate with audiences influenced by terrorists or the Russians. A stiff white guy in a suit, a general on a press junket, a Twitter handle sporting the American flag—these don't elicit sympathy *or* fear among U.S. adversaries. The solution during the counterterrorism years became "moderate voices": Muslim clerics preaching a brand of Islam that promotes more peace than violence. The idea was to amplify moderate voices in online discussions as a way to steer radical Islam back to the mainstream and dissuade vulnerable minds from going toward the darkness of terrorism. This ignored research showing that extreme positions outplay moderate ones and that negative content routinely outperforms positive content. Not only did moderate voices in the Arab world not perform well, but as soon as their connections to or support from America surfaced, their credibility among the diaspora audiences the West sought to influence dropped precipitously. To top it all off, American politicians who believe Islam itself is the root cause of terrorism bristled at the idea of backing any cleric.

America's messenger problem gets even more complicated when we're fighting Russian disinformation. Cold War U.S. information operations countering the Soviet Union employed American culture and art. Rock music blared from Voice of America, and Western news and television shows were piped past the Iron Curtain wherever possible. The messengers for America were its musicians, artists, and icons as much as its diplomats. The United States didn't talk about American values; it showed them by being American. In social media, Russia has nimbly made the messenger for its modern active measures appear American—false personas that talk like and look like Americans and help recruit unwitting Americans to the Kremlin's message. These false messengers attack, dispute, and degrade elected American officials who are the messengers for democracy against authoritarians. They've helped disable the credible voices America needs to preserve its system.

Since the presidential election of 2016, a murky cyber group known as the Shadow Brokers has released stolen NSA hacking tools into the wild. Release of these cyber weapons occurred in parallel with the release of materials pilfered from the CIA's Center for Cyber Intelligence. Their tools appear on WikiLeaks in batches, under the names Vault7 and Vault8. In combination, these hacking tools have empowered criminals, hackers, and even nation-states worldwide to shut down hospitals and steal millions from the accounts of countries and corporations. The open release of these hacking tools—by suspected Russian intelligence operators or a Russian organized crime syndicate—has further hurt the capability of the NSA, but that's nothing compared with the reputational damage to America. The Shadow Brokers pair their releases of cyber weapons dating back to August 2016 with social media taunts and public shaming on Twitter. Their tweets poke at the United States: "Is NSA chasing shadowses?" They also play to President Trump: "Don't Forget Your Base," "The-ShadowBrokers is wanting to see you succeed. . . . TheShadowBrokers is wanting America to be great again." Again, the United States suffers from catastrophic hacking that fuels a devastating influence campaign, harms the U.S. image, and damages the confidence of U.S. allies.

The Shadow Brokers and Russia's embassy in London have rightly pursued a human quality in their social media influence that American counternarrative campaigns just don't seem to get or fear employing. Major Chirchir, the Kenya Defense Forces officer who notoriously refuted al-Shabaab on Twitter, wasn't perfect in challenging terrorists, but he was far more effective, because he embodied the spirit and soul of a real human. Chirchir's account isn't a stuffy public affairs social media account posting scripted responses to current incidents or stale slogans with no meat.

Whether it's al-Shabaab, Major Chirchir, or fake Russian accounts meant to look like Americans, successful social media influence messengers must be human to be social, nimble to be effective, timely to be relevant, and adaptable to be successful. This is the appeal of President Donald Trump, whose personal Twitter posts, whether one likes them or hates them, effectively engage, enrage, and mobilize audiences. Sadly, no other American government leader or representative is given the autonomy to engage without scrutiny and joust without doubt the way the president does from his private platform. America's official social media messengers are neutered from the outset.

Beyond challenges with the message and the messenger, U.S. counterpropaganda struggles to seamlessly integrate all media in the way its adversaries do. The Islamic State, at its height, employed an entire media battalion that created, hosted, and distributed social media content at an unending pace. Twitter, Facebook, Telegram—the platform didn't matter; the Islamic State surfaced if an opportunity arose. Russia explores and exploits the entire range of social media to perfectly blend its influence efforts into a dominating information package. Its social media operatives use sites like LinkedIn and Facebook for reconnaissance of target audiences, gaining their preferences and relationships. Russian state-sponsored news outlets, particularly RT, have successfully hosted and shared YouTube content that reaches audiences they could never engage with traditional television. With each click and share of their propaganda, they gain cookies, internet traffic data, and even distribution-list sign-ups for further targeting of specific audiences. Forgeries maligning Americans and their interests can be placed in an anonymous site like 4chan or Reddit and then shared across the entire information spectrum. Twitter—for Russia or any other influence effort—provides the single best way to propagate a message around the world. Finally, the Kremlin saturates American audiences with freedom-loving American personas on Facebook, Instagram, or even Pinterest, inundating, on a person-to-person level, key accounts, mobilizing mavens among like-minded audiences. Russian propagandists get to operate all social media seamlessly, without scrutiny, legal rulings, or bureaucratic boundaries.

U.S. counterpropaganda efforts, depending on the agency, often treat these media as independent battle spaces. One organization may run a radio broadcast, another television, a third group social media perhaps. Bureaucratic program managers argue over their digital turf, and even try to develop resource allocations around social media platforms. During one early conversation at the time social media was emerging, after the Iranian Green Movement's protests, a veteran information operations expert asked me, "How do I buy MySpace? How do I buy Twitter?" He didn't mean that he wanted to buy the entire social media platform; he was simply trying to figure how to access these media. I was a bit confused. "There's nothing to buy,"

I said. "You just need to log in; it's free." This guy was one of the military's best, by the way. He wasn't trying to do anything wrong; he'd just grown up in a different era, under different authorities using wildly different methods.

Even if America did figure out the message, messenger, and medium it wanted to use to counter the Islamic State, Russia, or whatever comes next on social media, the method by which it conducts its influence business would fell even a perfectly designed effort. In the past decade, the government has not been able to figure out who will make the sausage and how it will be made.

\\\\\\\\\\\\\\\\\\\\\\\\\\\\

After I landed at Washington's Reagan National Airport, a short cab ride delivered me to a nondescript office building draped along the Potomac River. "We heard from some people you worked with at the Combating Terrorism Center that you were good at organizing counterterrorism research projects," said a typical defense contractor bureaucrat, between self-aggrandizing anecdotal stories of his military bravery and cunning brilliance as a former military officer. Apparently, this retired military officer, who'll be known hereafter as R.M.O., had been the mastermind behind all of America's psychological and information operations since the end of Vietnam. Yet, through the injustice of the military personnel system, he was now perched here in Reston/Rosslyn/Crystal City/Pentagon City/Bethesda (just pick your favorite one; they're all the same; the only difference is the commuting time). But it was all for the best, because he "got fed up with all the bullshit bureaucracy" and now he was "out here in the private sector," where he could "make more money and have freedom to do all the things in government" that he "wanted to do but never could do."

As is typical with defense contractors, his company had won a job about ten months earlier to "leverage open-source information to understand the strategic thinking of violent extremist organizations," or some other contemporary Pentagon jargon for terrorism.

I heard the R.M.O. introduction often. I knew it, would listen, and then wait for my turn to speak: "Yes, I helped run some research teams at West Point, and while they were never of the scale you created while deployed, I think I can definitely help you out." There it is, social engineering hook number 1: play to his ego. *I can do this project, sir, but not nearly to the level of your mastery . . .*

R.M.O. replied, "Well, it sounds like you might be able to help us. I designed this project, but you know how it is—I don't really have the time I'd like to devote to this effort. I've got so many other responsibilities keeping this company afloat, with business development, proposal writing, security clearances . . ."

I introduced my next play. "Look, you don't have to explain it to me; I totally understand. These companies never have enough former military talent to go around, and that's where I think I can help augment the great start you've already gotten on this project. Can you tell me a little bit more about what you need?"

This was where I would reel him in. Next he was going to push blame for all the project's problems onto a young analyst who he "thought could run the show" despite having no previous management experience or really any work experience at all, likely just serving in the company for only a year or so. Then he'd tell me how the project started off well, until the young analyst put in charge of it wanted "guidance." Turns out the analyst wasn't sure what was expected and had actually studied European politics rather than Middle East history, and spoke French or Spanish rather than the needed Arabic.

R.M.O. would continue explaining how the analyst had steadily developed a "bad attitude" over about a six-month period as the project wore on. The analyst didn't keep track of the budget, failed to turn in required updates, and then interviewed with a competitor contracting company and took a job with them, "on almost no notice," for "more money, doing the same kind of work."

What this R.M.O. wasn't going to tell me was that he could have hired me or someone similar to me to run this project ten months earlier. He may have even included my résumé in the proposal for the project, to help win the contract. But after winning, they decided to go with the cheaper, in-house "younger analyst."

R.M.O. began the pitch. "What we need is for you to review this project and see what can be salvaged, and make sure that we deliver to the client what we promised. I'd do this myself, but, with it being proposal season, I've got to be fully engaged in securing next year's work so everyone here has a job." That's right, R.M.O. was a martyr, too, desperately applying his brilliance to paper, creating magic formulas to defeat al-Qaeda, writing project proposals for government contracts that would bring millions of dollars to the company and save these poor analysts in need of work.

"This sounds like a great opportunity, a real challenge," I said encouragingly, and within minutes R.M.O. would be guiding me back to the "analyst pen," where a group of supersmart recent college grads were now being held hostage by their laptops, playing a passive-aggressive game of office *Survivor*, sitting amid dismal beige walls, stationed in cubicles like prisoners.

These bright young analysts had come to D.C. to save the world. They wanted to serve their country and their egos, ultimately becoming a congressman, a noteworthy academic scholar, national security adviser to the president, or the next brilliant analyst at the CIA. But instead they were stuck in the purgatory of defense contracting, lacking recognition for their hard work, insecure about their own talents, and isolated despite being surrounded by a sea of peers.

What the young analysts did love was their pay, and also telling their friends how they couldn't talk about their work "because it's sensitive." The work wasn't as sensitive as their egos, though. Their daily grind consisted largely of advanced Googling, followed by paraphrasing of things written by others. Inside their cubicle farms, the isolation from their managers and their peers drove them insane. To get onto the project they wanted or be promoted before their peers, they would engage in the usual office backstabbing. They'd sabotage the smartest challenger in the office, generally sullying their own image in the process. The analyst pen brought out the worst in them and those sitting to their left and right.

I began listening to each of them as I made my way around the analyst pen. They described what they were working on, what they were trying to do. Each of them was smart and capable, but dying for direction. I offered some quick thoughts as I circled the pen with R.M.O. leering over my shoulder. He passed judgment on the analysts and imparted his knowledge and authority. Given only a little encouragement and a few comments, they seemed as if they'd been lifted up. *Has anyone ever talked to these people?* I thought to myself. In minutes, I'd become some sort of horse whisperer for intelligence analysts while doing absolutely nothing thus far.

As I navigated through the analyst pen, I realized that this project was due in a matter of days and the company didn't really have much of a plan. I also realized that I'd seen this project before. It was glaringly familiar. I pitched back to R.M.O., summarizing a way forward for the effort, and he gave me the nod.

"Clint, that sounds great—I'm glad you'll be able to help us, and if there is anything you need to get started, just let me know." R.M.O. was closing the deal. Nice. What I wanted . . .

Next I'd get the contract and start heading to the airport, and what R.M.O. didn't know was that I already knew how to do this project because I'd helped *invent* this project. A year before, I had worked for his predecessor, the other R.M.O. before him, who had requested me on short notice to write the technical portion of a government proposal that ultimately became this project. That's right, he didn't even know or remember—it's all a shell game, folks, a shell game of horseshit smuggling, and now I was returning to Reagan Airport to finish this project from a laptop computer perched at a coffee shop in your neighborhood.

In a month, an Adobe file and supporting PowerPoint slides would be emailed through a series of nondescript office buildings around the Washington, D.C., metropolitan area. One contracting company would pass it to the next, until finally an officer eating lunch in the Pentagon got it in his Microsoft Outlook inbox. The project wouldn't be groundbreaking, nor would it be awful; it would simply be sufficient for securing the defense contracting company's next contract, and then immediately forgotten by everyone who received it. The electronic files would blend in with the thousands of other counterterrorism projects circulating through the U.S. government. These projects are indistinguishable from the analysis of al-Qaeda and the Islamic State that you, the public, can discover on the internet, because more often than not they will be a summary of the internet. The project would go to the analysis graveyard, buried amid the hard drives of America's military/intelligence/homeland security bureaucracy.

In real life, this story didn't happen exactly this way. Have you ever watched Oliver Stone's movie *Platoon*, where seemingly everything awful in the history of the military happens to Charlie Sheen during his year in Vietnam? That's what I've done here: combined all of the worst aspects of my defense contracting time into a few pages. It's an amalgamation of dozens of encounters I've had with dozens of defense contractors over a dozen years supporting the military/intelligence/homeland security industrial complex. Each of these things happened, but in different places, with different R.M.O.s running pens of different disgruntled intelligence analysts. Some of these contracting companies are quite good—better than the government entities they support—but many are downright awful. And don't get me wrong: there are many great retired military officers doing exceptional work for their country. But this is a certain kind of character I frequently run into.

What I describe above is the worst-case scenario of a bad industry—specifically, the contracting of counterterrorism analysis to that part of the private sector known largely as "intelligence support to information operations." It's what happens when the characters from the movie *Office Space* get placed in the TV show *24*. It doesn't work. I don't know anything about other forms of defense contracting: weapons, maintenance, etc. The story above describes the circuitous spin of Microsoft PowerPoint slides and Word documents, funded by the American taxpayer, slowly coming together to inhibit rather than accelerate America's counternarratives programs. The bulk of U.S. influence funding, distributed through roughly a dozen different agencies and military commands, ultimately routes to a set of defense contracting companies largely staffed by people who once worked in the same dozen or so agencies and military commands that provide the funding.

I learned one thing from my time working in the big defense contract system supporting psychological operations, information operations, and public affairs operations—a group I collectively and affectionately refer to as the Men Who Stare at Goats, after the 2004 movie, starring George Clooney, that showcased military psychological operations. Despite all the conceptual complications America faces in fighting an information war, the largest impediment may be structural—who is in charge of the messaging and who will do the work.

During the Cold War, the U.S. Information Agency (USIA) served as America's principal arm devoted to public diplomacy with the express mission, "to understand, inform and influence foreign publics in promotion of the national interest, and to broaden the dialogue between Americans and U.S. institutions, and their counterparts abroad."[3] After the Soviet Union collapsed, the victorious United

States didn't have any foreign audiences that needed to be informed or influenced, or so they thought. Democracy and the West had won, and the USIA disbanded in 1999, placing its broadcasting functions into a new outfit called the Broadcasting Board of Governors and its non-broadcasting components into the new office of the under secretary of state for Public Diplomacy and Public Affairs.

As the USIA disbanded, al-Qaeda's global message gained steam. Just two years later a global Muslim minority called al-Qaeda waged war on the U.S. homeland. Major military offenses were launched in Afghanistan and Iraq and the big budgets went to the Department of Defense. Within the military, influence operations were divided into different strata, partitioned confusingly between public affairs, information operations, and psychological operations (PSYOP). When military units are deployed, they traditionally employed their PSYOP troops for handing out pamphlets, dropping fliers, and radio broadcasts. Information operations planners at the brigade levels and above supported military operations through information campaigns—deliberate strategies to raise awareness of issues and change the mindsets of specified populations—while public affairs officers representing generals and high-level DoD officials conducted press briefings. Their efforts at the start of the war on terror weren't bad, but they were small, limited in effect, oftentimes overlapping, and in the worst cases contradictory.

Operation Iraqi Freedom (OIF) was a quagmire that worsened each year, and the larger mission of counterterrorism in Afghanistan and around the world, Operation Enduring Freedom (OEF), ground on, such that by 2005, 2006, and 2007, a new counterinsurgency approach pushing "whole-of-government" solutions—a combination of actions pursued by many agencies with shared objectives—became the mantra for winning intractable wars and bringing U.S. troops home. The United States sought to win "hearts and minds" on the battlefield and online, and "strategic communications" became the catchall term for how the U.S. State Department and the Department of Defense would come together to influence the world's populations susceptible to jihadi recruitment. Buckets of money—no, let me clarify that: truckloads of money—were spent across the entire spectrum of influence campaigns, doled out to military combatant commands and the State Department through convoluted campaign funds.

America recognized that it was in a new and different information war but didn't know entirely what to do or who should do it. The military's PSYOP troops, designed for traditional media, weren't well equipped for digital domains. Their officers were retrained from more traditional military specialties like infantry or artillery. The State Department, throughout its diplomatic history, had more know-how, but few resources. The intelligence community had experience from the Cold War, but it didn't have the authority to meddle in cyberspace. All feared violating laws forbidding the U.S. government from influencing U.S. populations—an increasingly impossible task when engaging social media audiences not defined by international borders. The agencies needed to integrate and synchronize to be successful, and by 2006 the Bush administration knew a central point must lead influence efforts.

The United States' push for more "soft power," the kind Secretary of Defense Gates spoke of, began in earnest in 2006 with the launch of the Counterterrorism Communication Center (CTCC) to coordinate the morass of contradictory messaging coming from the United States Departments of State and Defense. This began a string of mutations every few years as U.S. agencies, vying for funding, wrestled for control over messaging and messages. The Global Strategic Engagement Center (GSEC) replaced the CTCC in 2008, and the Center for Strategic Counterterrorism Communication (CSCC) then replaced the GSEC in 2011. Each manifestation signaled the failure of the previous version to achieve meaningful gains or survive bureaucratic infighting. The CSCC incarnation arrived just as terrorists were aggressively moving to social media. Underfunded and overmanaged, the CSCC battled its overlords on the National Security Council and its funders in Congress as much as it fought al-Qaeda overseas. By 2012 and 2013, some good successes had come from its Arabic-language YouTube videos, which drew in foreign audiences. These CSCC placements showed promise for two reasons: some of the government's best and brightest created them, and the messages didn't just try to champion democracy, but directly and intelligently refuted terrorists' messages. Because the CSCC communicated in a foreign language most Americans didn't understand, they were relatively free to experiment without fear.

Then came al-Shabaab. As @HSMPress and Kenya's Major Chirchir battled on Twitter, Somali American men from Minneapolis streamed into Somalia. The mission to counter al-Shabaab in the English language on Twitter went to the CSCC, signaling the slow demise of the center. The CSCC resisted tweeting in English. As one of its savvy officials confided, "The moment we do it [tweet in English], the eye of Sauron will gaze upon us," an allusion to the fiery eye in *The Lord of the Rings* that turned to look when its master's name was invoked. The officials' foresight proved correct.

Tweeting in English under the @ThinkAgain_DOS Twitter handle attracted the attention of the entire State Department and the entire Washington media circus, which doomed the CSCC. Those inside government questioned every tweet. The media ridiculed CSCC responses. Terrorists and their sympathizers, unencumbered by approval processes and bosses, jabbed back at the State Department's account in the open. A persona named Abu Ottoman tweeted back to @ThinkAgain_DOS, "Your boss is going to fire you soon if these tweets don't improve."[4] Even the terrorists knew that the CSCC experiment wouldn't last long. State Department analysts manning official accounts couldn't win, boxed in as they were by bureaucrats, not allowed to think nimbly or respond quickly, heckled by the media and terrorists. @ThinkAgain_DOS and the CSCC dissolved like its predecessors, absorbed into another organizational restructuring and acronym: the Global Engagement Center (GEC).

In 2016, the GEC sought to reboot America's counterpropaganda with new blood and a new approach. Under its new leader, former Navy SEAL Michael Lumpkin, the GEC wanted outsiders from "Silicon Valley and Madison Avenue who can help us work through our approach and the information battle space."[5] The idea wasn't new per se. The information operations groups of the military and the public diplomacy wonks of the interagency had flirted with technologists before. Again, challenges quickly emerged on both sides in trying to blend these communities. Silicon Valley approaches focused largely on big-data analytics and convincing people to buy things, whereas the U.S. government sought to quell the rhetoric of extremist bad apples and sell them a new idea. The government wanted terrorists to change their behavior in a positive way, which was a far deeper challenge than moving products from retailers to social media users. Even if the GEC could recruit whiz kids from the coasts into the dungeons of government bureaucracy, they'd be limited by their lack of security clearances, less effective tools, and tight oversight.

The GEC expanded its mission in late 2016. The Countering Foreign Propaganda and Disinformation Act, introduced by Senators Rob Portman and Chris Murphy, increased the GEC's scope and funding, but as of this writing, the Trump administration's State Department has yet to move against Russian social media influence in a deliberate way. The GEC sits in idle a year after the presidential election, seemingly uncertain about how to move forward.

The organizational challenges hampering the government weren't limited to who was in charge; there was also the question of who would do the work. Whereas the military's original PSYOP troops owned, rented, and operated their own printing presses and radio operations, no government entity maintained the technology and tools necessary to engage in social media. Even if a government agency possessed the tools, no human resources in the U.S. government hosted the talent to create engaging social media content at the volume needed. That's where defense contracting comes in.

Defense secretary Donald Rumsfeld's lighter-footprint approach to the invasion and occupation of Iraq depended on the nimble employment of contracted services. Contracted services for everything but frontline troops sought to scale America's battlefield mobilization and sustain war without increasing the size of the military or building out highly specialized units that wouldn't be needed during peacetime. The costs were projected to be cheaper over the long run, and the approach provided flexibility to the government. Public diplomacy, information warfare, psychological warfare—all highly specialized disciplines— seemed perfect for employing defense contracting on a temporary basis. The government, through contracts, could rent highly trained specialists and buy specialized tools. That all makes sense on paper, if one doesn't really understand how defense contracting works.

Defense contracting during the global war on terror and even today occasionally works, but when it comes to influence operations, it's largely been a disaster. Defense or intelligence community funding has been directed less toward getting the right talent or technology and more toward sustaining the flow of funds to prime vendors, largely staffed by former military and intelligence officers. The system sustains itself by imposing two barriers to competitors: contract vehicles and security clearances. Contract vehicles, in concept, seek to validate a vendor for selling to the government. They are said to establish predetermined prices for products and services, streamlining government purchases. In reality, it's a deliberate bureaucratic maneuver that requires a vendor to go through costly applications and approvals just to offer services to the government. The hurdles for simply selling to the government get so high that small vendors, those with specialized talent and innovative tools, can't compete with the big contractors that dominate the market, those business behemoths that can afford to feed the government's bureaucracy.

Security clearances are another barrier for influence operations talent acquisition. The dawn of counterterrorism created legislation and funding that required personnel to maintain security clearances to qualify for funding, suggesting the need to keep secrets. Defense contractors loved the mandate of security clearances, because they narrowed the pool of talent to their own employees and favored their large companies, which could bear the costs of background checks and maintaining security protocols. For the military and the intelligence community, security clearance mandates in contracts served two purposes. They ensured that outside contractors would be vetted and screened, most always returning contractors looking like them and talking like them. And mandating security clearances for defense contracting meant more jobs for retiring military officers and intelligence professionals. They could retire from government service and, based only on their security clearance, be nearly guaranteed a job. The result is that the defense contracting system creates a mirror image of the government, and few fresh ideas or talented social media analysts can penetrate the system to support the government.

The government organizational amoeba and the contracting system that supports American counterinfluence from afar costs billions of dollars to run and sustain, and it's still not as effective at mobilizing audiences as its adversaries' low-budget operations. The Islamic State's media battalion, at its height, numbered a few dozen people in Syria and Iraq, with some voluntary surrogates stationed around the world. They produced posts nearly nonstop and hosted new, engaging, high-quality social media videos several times a week. Their talent was mostly self-taught: volunteers with a knack for graphic design and video editing. Russia's Internet Research Agency produced and presumably still runs around-the-clock social media influence campaigns in multiple countries. The few former employees who've spoken of their time working there suggest that there are roughly two hundred people operating for the Russians globally. The height of the U.S. election may have seen only ninety people influencing the U.S. audience on social media. Their operations span three floors, dividing and compartmentalizing bloggers, social media operators, and meme makers. The Russians meddling with Americans, Europeans,

Ukrainians, Turks, and Syrians are computer whizzes and journalists, young contractors hired by the hour for a specific purpose, interchangeable widgets in a simple yet effective media machine.

By contrast, the U.S. government easily employs thousands of people, either inside government agencies or via outside contractors, to counter terrorist influence operations. Many plans are made, few are executed, and what little content has been developed and deployed by the U.S. government bureaucracy has failed to make a dent in terrorists' efforts. Many American counterinfluence personnel have limited knowledge of social media or have never even used it, either too old to know it or unable to access it from inside government firewalls. Even when military or State Department groups doing counternarratives acquire good people, which they do (the smartest person I know in the influence profession works inside the government), they receive tremendous oversight and little autonomy.

The terrorists and the Russians conduct their operations openly, without barriers to the internet or constraints on their actions, and thus those Westerners best positioned to analyze them and fight them have not been trapped behind the gates of intelligence compounds or buried behind the walls of the Pentagon. They spend their days on social media, analyzing terrorist videos or Russian propaganda. Many of them seek to fight terrorists or protect democracy from Russian interference, but they'll never even get in the door. They don't have clearances or would never get one, having smoked marijuana too many times or failed to acquire an unnecessary prerequisite degree in an unrelated topic. Even if they do get in the door, they won't last long. Their well-crafted social media campaigns would never clear the bureaucratic obstacles for launch or pass the scrutiny of gatekeepers and stakeholders.

I can counter terrorists or Russia from home better than I could from inside government, and so could most of the great civil servants America has working on counternarratives. I'm not smarter than any of the government folks doing counterinfluence or many of the contractors spread around D.C.'s Beltway. I just have more flexibility, can practice, and am free to fail at my house without scrutiny. There's no barrier to entry for me as a citizen. I can make a thousand mistakes; I can learn from others and plod away without a manager telling me to stop or change something to make a politician happy.

In the end, the most effective way to counter terrorist influence has been to simply kill terrorists. Despite all the calls for American counter-messaging, when terror groups wane, their influence evaporates. As the Islamic State's remnants now run scurrying from Iraq and Syria, the appeal of their social media has declined and the desire for new-comers to join has subsided a bit. Going forward, though, the United States' and really the West's dilemma will be how to fight authoritarian regimes like Russia on the social media battlefield. The response to foreign meddling in America will be far more challenging, intricate, and demanding.

Some solutions appear easy. The FBI, when conducting its investigations into hacks, should anticipate how what was stolen might be used for the purposes of influence. Anticipating rather than reacting to *kompromat* can help inoculate the victim from negative influence. The Department of Homeland Security should provide public responses to falsehoods and smears launched by foreign adversaries regarding U.S. domestic issues. The State Department should do the same in response to falsehoods related to U.S. foreign policy. The Department of Defense and the intelligence community must rapidly create systems for tracking Russian social media disinformation to anticipate and ultimately counter the Kremlin's march. Finally, the West collectively must decide how to respond to Putin's manipulation. Employing the Kremlin's own approach against the Russian population by meddling with Russia's election is not the answer. Doing unto Russia what Russia did to America would only erode American values and undermine our credibility. Countering Russia requires a defense of our country and the promotion of our beliefs, and this is where America currently lacks the will and ability to defend itself: leadership.

Ultimately, America's problem in counterinfluence is that we don't know what to say, because we don't know what we believe in. In war and peace, for more than two centuries, America's elected leaders have unified the country and its strategy against adversaries by uniting Americans toward common goals. During the Cold War, the United States promoted democracy and democratic values. But today the United States doesn't appear to know what it wants. Quite simply, if America doesn't have its feet on the ground, then it can't punch back at those challenging us. Leaders in both political and civil society must clarify what America believes in.

The United States might also consider why its image advanced globally in previous eras, but not today. America didn't project its values around the world through catchy promotional advertisements and public affairs messaging. The United States accelerated past all other countries by being great, not telling everyone how great it was. Art, music, science, free markets, natural disaster relief—the United States did things no other country would, and accomplished things no other country imagined or even could. U.S. policy for influence operations might ultimately be to do nothing at all, except figure out what we stand for, what we believe in, and what we will again fight for. The answers to those questions will ultimately be the nation's counterinfluence message. The strategy will reveal itself. When Americans are the best versions of themselves, influence happens naturally. Sadly, the United States and many parts of the West have pulled back from the world stage in recent years, and until we have a better message, social media propaganda machines will have their way with American minds. And it won't be the Russians who come to dismantle our collective psyche; it will be Americans willfully doing it to one another, in pursuit of their own personal gain.

9

From Preference Bubbles to Social Inception: The Future of Influence

As a kid growing up in Missouri, I'd occasionally encounter ham radio operators—communication enthusiasts, one might say—shooting signals into the stratosphere simply to see if it was possible to communicate with someone on the other side of the world. Humans love connection, and the internet opened a door to a world of connections that ham radio operators could never have imagined. Regardless of geography, people who thought alike and had the same interests, pursuits, dreams, and goals could converse, create, and confide.

Internet evangelists saw the world in a positive way as we entered the new millennium in 2000. The World Wide Web broke the boundaries of the local and opened up a global community. The personal computer, now connected to the internet and search engines like Netscape and Google, brought any interest, fact, or item to the fingertips of those connected to the web. Physical boundaries no longer mattered, and those passionate about chess, cancer research, or the television show *Friends* could find like-minded enthusiasts around the world wanting to share their thoughts and experiences. Those previously under oppressive regimes, denied access to information and the outside world, could leverage the web's anonymity to build connections and share their experiences and hope for a better world, either at home or elsewhere.

The best intentions and perceived outcomes shine bright with each advancement in information technology. The internet's open system led to niche repositories for highly specialized information appealing to smaller audiences distributed around the world. Chris Anderson of *Wired* magazine famously detailed this phenomenon from a business perspective in his book *The Long Tail: Why the Future of Business Is Selling Less of More*. Building from the research of Erik Brynjolfsson, Yu (Jeffrey) Hu, and Michael D. Smith, Anderson explained how online access creates not just lower prices but increased product variety. In the pre-internet era, where traditional local markets offered only a small range of high-selling goods, the World Wide Web offered an opportunity for things like books, music, and homemade goods to be sold at lower volumes over an extended period. The "long tail" referred to a high-frequency power distribution. Quite simply, the internet made it possible for those on the fringe to sell their products to larger audiences over longer time periods, because there were no costs to keeping products on the market. The audience would eventually find what they wanted somewhere on the internet.

The long tail didn't just apply to products, but also to ideas. Remote-controlled airplanes, cross-stitch, fantasy football, brewing beer at home—somewhere in the world, someone was interested in the same topics, concepts, or even hatreds as someone else in the world. The internet removed the barriers between these people and dramatically lowered the costs of communication and, later, of the production of content.

The upsides were apparent, but in the euphoria of any new advancement, few calculated the downsides of anyone in the world being connected to anyone else in the world. Hackers and cybercriminals were some of the first actors to exploit the internet in pursuit of money and fame. They worked alone in the beginning, but over time they came together in their illicit pursuits. Beyond the technical trickery of hacking, hate groups and terrorists found the internet an anonymous playground for connecting with like-minded people. There were only a handful of extremists, or possibly only one, in any given town, but with the long tail of the internet, there were now hundreds and even thousands of extremists who used online congregations to facilitate the physical massing of terrorists in global safe havens or remote compounds.

Think back to jihadi militancy before al-Qaeda. It took the mujahideen a decade to raise awareness, radicalize, and recruit merely a few thousand international volunteers to fight in Afghanistan against the Soviet Union. Cassette tapes, paper fliers, and rotating proselytizers roaming the earth connected with and brought in sympathetic supporters interested in pursuing violent jihad. Al-Qaeda—the Base—trained small groups to deploy as global evangelists for its cause among any and all wars in Islamic countries. Bin Laden's indoctrination process certified fighters for his brand. A training camp graduate was an expert in both mind—ideologically adherent to militant jihad—and body: physically skilled and ready to conduct combat operations or terrorist attacks. In Somalia, Sudan, Yemen, and Afghanistan, al-Qaeda's reach before the internet remained limited to a few loyal, trained, and indoctrinated supporters who preached bin Laden's vision and pursued his violent goals.

The internet provided a virtual safe haven for bin Laden's al-Qaeda, allowing the small minority of Muslims inclined to jihadi extremism to connect with like-minded supporters in the "long tail." Those thinking like bin Laden, or wanting to be like bin Laden, could hear his words, see his group's attacks, and connect with his organization. As counterterrorists scoured the earth searching for al-Qaeda's head shed, the internet provided enough cover, capacity, and space for the terror group to survive physically by thriving virtually. This made al-Qaeda bigger, but not necessarily better—more diffuse and elusive, but vulnerable to fissures and difficult to manage.

The internet brought the world together, but, over time, social media has torn the world apart. There are many reasons why this has happened, but one factor stands above the rest: preference. Unbridled preference—man's ability to make nearly endless selections on social media—when accumulated on a global scale, has torn the fabric of societies, crippled democratic institutions, and polarized audiences into virtual and physical bubbles. The relentless pursuit of preferences turns smart crowds into dumb mobs, leads to the selection of preferred fictions over actual facts, and creates an environment where humans have access to more information than ever but actually understand less about the physical world. The power of preference now haunts not only al-Qaeda as they've been outpaced by the social-media-savvy Islamic State, but America, the land that created social media. No barriers to entry and unlimited preference in the virtual world have overtaken compromise in the real world. Online, the pursuit of comfort and confirmation create an alternative reality.

Those researching the internet and social media didn't expect this deterioration, and even the social media companies are just now beginning to understand what's happening. In *The Long Tail*, Chris Anderson predicted that preference would lead to bliss. The heading of one of his chapters was "The Paradise of Choice: We are entering an era of unprecedented choice. And that's a good thing." Anderson was right about the internet bringing us together, but his optimism about preference has not been borne out. Unlimited variety and choice have somehow divided us, made us angry for reasons we can't explain and bitter toward our own countrymen, friends, and family who don't share our preferences.

Similarly, Andrew McAfee and Erik Brynjolfsson, in their 2017 book *Machine, Platform, Crowd*, describe how modern crowds, empowered by smartphone technology, now contribute to the knowledge of society and can work together to identify solutions. They explain how, before barriers were lowered with the internet, societal elites and their institutions maintained collections of knowledge that they used to power their products and further their agendas. Trained professionals selected, arranged, and maintained information collections representing what they describe as a "core"—a set of dominant organizations, institutions, groups, and processes. The dawn of the internet, further accelerated by smartphones and social media, has generated an alternative they describe as the "crowd"—a user-generated library of free information creating vast amounts of knowledge at a breakneck pace. They assert that this online crowd now outperforms the core, upending businesses and entire industries and society itself.

Up to the dawn of social media and in its early years, the crowd did outperform the core, powering unprecedented economic change and altering the way societies and governments communicate. The right crowds are definitely smarter than the core. But we often forget, as I did when crowdsourcing terrorism analysis, that the "core," the technological and academic elite, dominated the internet and were the first to arrive on social media. The initial members of the crowd were those privileged enough to have internet access and afford a smartphone. An early crowd member—from the mid-1990s until the late 2000s—needed some education and training to create and moderate content on forums, blogs, and chatrooms. Mark Zuckerberg started Facebook at Harvard, after all, not in the failed state of Somalia. The first crowds on the internet and social media were more educated, experienced, and privileged—collectively smarter than the core. The experts at the core still held a repository of experience, reasoning, and knowledge to effectively harness the crowd's energy for discrete tasks and specified disciplines, determining what insights and innovations were outpacing existing pre-internet libraries and industry practices. They were able to judge the merit of new discoveries and employ them. But today's crowds are anyone and everyone with a cell phone and a Facebook account, very different from the limited and much smaller virtual crowds of only a decade ago.

My experiences with the crowd—watching the mobs that toppled dictators during the Arab Spring, the hordes that joined ISIS, the counterterrorism punditry that missed the rise of ISIS, and the political swarms duped by Russia in the 2016 presidential election—lead me to believe that crowds are increasingly dumb, driven by ideology, desire, ambition, fear, and hatred, or what might collectively be referred to as "preferences."

Eli Pariser, the head of the viral content website Upworthy, noted in his book *The Filter Bubble* the emergence and danger of social media and internet search engine algorithms selectively feeding users information designed to suit their preferences. Over time, these "filter bubbles" create echo chambers, blocking out alternative viewpoints and facts that don't conform to the cultural and ideological preferences of users.

Pariser recognized that filter bubbles would create "the impression that our narrow self-interest is all that exists."[1] The internet brought people together, but social media preferences have now driven people apart through the creation of *preference bubbles*—the next extension of Pariser's filter bubbles. Preference bubbles result not only from social media algorithms feeding people more of what they want, but also people choosing more of what they like in the virtual world, leading to physical changes in the real world. In sum, our social media tails in the virtual world wag our dog in the real world. Preference bubbles arise subtly from three converging biases that collectively and powerfully herd like-minded people and harden their views as hundreds and thousands of retweets, likes, and clicks aggregate an audience's preferences.

When users see information that confirms their beliefs, they like it, share it, and discuss it. Social media subtly creates large-scale confirmation bias—the tendency to search for or interpret information in a way that confirms previously existing beliefs or preferences. When users see information that challenges their beliefs and desires, they may block the disseminator of the content. Or if they want to challenge the disseminator, they'll seek out and promote content directly disputing it. The information employed to settle such disputes may or may not be true, but what is important is the validation received by the person pushing it. Confirmation bias gets worse when users get emotional, triggering their instinctive fight-or-flight tendencies. Competitions, disasters, and defense of our core values make users reach for this preferred information that helps them feel secure.

Social media amplifies confirmation bias through the sheer volume of content provided, assessed, and shared. But social media also connects users to their friends, family, and neighbors—all people who, more often than not, think like they do, speak like they do, and look like they do. Social media users see news, information, and experiences contributed by their friends and followers. They naturally tend to believe this information as a result of implicit bias—the tendency to trust people we consider members of our own group more than the information of an outside group. Users trust the sender and transitively trust the information being sent, regardless of whether it's accurate or not.

Confirmation bias and implicit bias working together pull social media users into digital tribes. Individuals sacrifice their individual responsibility and initiative to the strongest voices in their preferred crowd. The digital tribe makes collective decisions based on groupthink, blocking out alternative viewpoints, new information, and ideas. Digital tribes stratify over time into political, social, religious, ethnic, and economic enclaves. Status quo bias, a preference for the current state of affairs over a change, sets into these digital tribes, such that members must mute dissent or face expulsion from the group. Confirmation, implicit, and status quo bias, on a grand social media scale, harden preference bubbles. These three world-changing phenomena build upon one another to power the disruptive current bringing about the Islamic State and now shaking Western democracies.

\\\\\\\\\\\\\\\\\\\\\\\\\\

"I wasn't the best because I killed quickly. I was the best because the crowd loved me. Win the crowd and you'll win your freedom." So declared Antonius Proximo, the fictional mentor to Russell Crowe's Maximus in *Gladiator*. Maximus responded, "I will win the crowd. I will give them something they have never seen before." Proximo understood crowds and mobs, and he would have dominated today's social media. *Clickbait populism*—the promotion of popular content, opinions, and the personas that voice them—now sets the agenda and establishes the parameters for terrorism, governance, policy direction, and our future. Audiences collectively like and retweet that which conforms to their preferences. To win the crowd, leaders, candidates, and companies must play to these collective preferences. Leaders aren't born or made today; they are accrued through the sum of clicks on iPhones and searches on Google. Those who give the crowd what they want will gain what they need to advance their agenda. The more a person plays to the crowd's preferences, the more they will be promoted and the more power they will gain. Power, once attained, can then be used to issue preferences back to the crowd, as the crowd will want to champion and sustain that which it has created, whether it be the caliph or the president.

Clickbait populism drives another critical emerging current: *social media nationalism*. Tribes and clans for centuries have been delineated by common descent, history, culture, or language, as defined by the physical territories they inhabit. Flags, symbols, slogans, speech, dress, and traditions characterized nations in the real world, mobilizing societies under the banners of shared values and common enemies. But with the advent of the internet, many people now spend more time with others online than in person, connecting with like-minded virtual personas who increasingly reside outside the physical terrain where they live their daily lives. Each year, as social media access increases and virtual bonds accelerate, digital nations increasingly form around online communities where individual users have shared preferences. Social media users now reside in virtual social media nations, where they identify their allegiance through the employment of hashtags, similar biographies, and symbolic photos. These virtual cues allow users to connect with members of their social media nation, feel reinforcement for their beliefs, and shape collective digital values, regardless of whether they exhibit any of the virtual nation's principles or behaviors in real life. Social media nations (i.e., competing preference bubbles) fight in persistent information wars on digital battlefields—social media platforms—as they advance their nation, real or imagined, against their digital adversaries. This social media nationalism provides its virtual citizens with shelter from reality and replaces real-world compromise with the collective virtual pursuit of shared preferences. The more people live in the online world, the more their online world defines who they are. As places increasingly come to look similar through corporate franchising and cookie cutter communities, people may be inclined to identify as being from the #TrumpTrain or the #Resistance rather than from Texas or New York. Which social media nation someone identifies with provides more information about someone than the state or country they are from.

Social media nationalism and clickbait populism have led to a third phenomenon that undermines the intelligence of crowds, threatening the advancement of humanity and the unity of democracies: *the death of expertise.* As the barriers to internet access got lower and lower, anyone, regardless of education, training, or status, could explore information and voice their opinion in debate. This would seem, on the surface, to be good for democracies, as increased information, awareness, and voice would seem to encourage more civic engagement and debate and better collective outcomes. Instead, social media users, in their relentless pursuit of preferences, have selectively chosen information and expertise they like over that which is true or even real. Social media users participating in the crowd have chosen to be happier and dumber by not just challenging McAfee and Brynjolfsson's core but also by seeking to destroy it.

Tom Nichols, a professor at the U.S. Naval War College, articulates the third phenomenon in his 2017 book *The Death of Expertise*. Nichols describes how the internet opened the doors of information to anyone, and, through social media, users have cherry-picked data and opinions to support their preferred belief. He argues that in the modern era, "any assertion of expertise produces an explosion of anger from certain quarters of the American public, who immediately complain that such claims are nothing more than fallacious 'appeals to authority,' sure signs of dreadful 'elitism,' and an obvious effort to use credentials to stifle dialogue required by 'real' democracy."[2] Social media users, in their relentless pursuit of preference, believe that, with only a few clicks on their phone and a Google search or two, they are as smart as or smarter than a medical doctor, a college history professor, a physicist, a dietitian, or their elected leaders.

Social media makes everyone believe they are an expert on everything, highly informed because they can access so much information, and verifiably smart based on their aggregation of a large number of friends, followers, views, retweets, and likes. Today's social media nations seek not only to challenge the core but to rewrite history, quash science, and alter governance to conform to their preference bubble. Clickbait populism, social media nationalism, and the death of expertise have created two shocking preference bubbles that, within only three years, have upended international security and the world order: the #IslamicState and the #TrumpTrain.

The Islamic State of Iraq's rebirth and ultimate triumph in establishing a state resulted from playing to its supporters' preferences. For social media audiences, the modern Islamic State cherry-picked bits of Islamic history to create the justifications for traveling to Syria and Iraq. This preferred apocalyptic narrative led zealous foreign fighters, from Europe to the Philippines, to die in droves in pointless battles such as that of Kobani, Syria, in 2015—a historically significant town in the Koran but a futile military effort on the modern battlefield. The Islamic State never truly governed in line with the days and times of the Prophet Muhammad, but that didn't stop tens of thousands of men and women, entire families in some cases, from traveling to a war zone to pursue this social media fallacy.

Donald Trump, the descendant of a wealthy real estate developer from New York City, drew support from less educated, poorer Americans. Trump's magic formula, his understanding of how to win, came not from any experience in governance, nor even a great track record in business (he's declared bankruptcy several times), but from his understanding and mastery of self-made marketing via reality television. Reality television's formula is simple: keep people watching by giving them what they want, an endless soap opera, sustained drama, constant conflict, heroes, villains, and storylines that are easy to follow.

His social media pontifications promote topics and themes that play to his audience's preferences more than they describe actual policies or reality. The online herds that supported, promoted, and now sustain President Trump shared certain preferences: anti-immigration, anti-Muslim, nationalist rather than globalist, toughness in general but particularly against the Islamic State, low taxes, repeal of Obamacare, and even standing for the national anthem rather than kneeling. The more a narrative grows in popularity, the more Trump amplifies it. The more he champions what the audience likes, the more support he accrues from this preference bubble. Trump wins the crowd regardless of whether he wins anything *for* the crowd.

Trump's most popular themes leading up to the election were "emails," "Lock her up," and "Build that wall." At rallies and on Twitter, Trump yelled about the investigation into Hillary Clinton's emails, but none of his social media posts mentioned that the investigation ended twice, and that there were many varieties of emails— those involved in the Clinton investigation, those hacked from the DNC, others pilfered from John Podesta, former secretary of state Colin Powell, General Breedlove, those discovered in the Anthony

Weiner investigation. Trump's calls to arrest his political opponent continue to bring cheers from the crowd, but his supporters may not entirely know why she should be detained. Trump's promise of a U.S. border wall—and, further, that Mexico will pay for it—by his own admission, in a privately disclosed phone call, represents good politics but bad policy. He is an expert at clickbait populist narratives, and repeats them whether or not they make sense or he actually intends to pursue them.

Social media nationalism among the #IslamicState and the #TrumpTrain looks strikingly similar. A leading indicator of whether someone supports or is a member of al-Qaeda or the Islamic State has been the use of Osama bin Laden, Anwar al-Awlaki, or jihadist black flags as a Twitter or Facebook profile picture. As early as 2012, a simple count of Facebook avatars hosting bin Laden/Zawahiri images, versus Zarqawi or ISIS flags, provided a general sense of the shift between al-Qaeda and the emerging Islamic State. Jihadi supporters post scriptures or quotes from their favorite leaders in their biographies. Each seeks to identify with a movement, so much so that J. M. Berger was able to conduct an online census of jihadis in 2015. Using only openly available information from Twitter, Berger showed growth in ISIS's support, estimated which countries hosted the most online supporters of the Islamic State, noted the top hashtags signifying Islamic State support, and even detailed the communication patterns of those most sympathetic to the Islamic State. The Islamic State's social media nationalism could thus be seen and measured.

Social media nationalism for President Trump is more pronounced, overt, and apparent to anyone following him on Twitter. Avatars showing red hats emblazoned with the slogan MAKE AMERICA GREAT AGAIN are accompanied by the corresponding hashtag #MAGA. The social media biographies of Trump supporters self-identify with their nation, denoting themselves as Christian, conservative, or a proud citizen of their state, adding in the hashtags #BuildTheWall, #Emails, #LockHerUp, #GOP, and #Trump. Trump's opposition employs the same tactics, his detractors deploying countervailing hashtags like #Resist, #ImpeachTrump, and #NotMyPresident. These social media nations share common information and opinions in their preference bubble to promote an ideology or person, real or imagined, above those in opposition to it.

The death of al-Qaeda's expertise due to clickbait populism was evident years before bin Laden's death. The English-speaking Yemeni American cleric Anwar al-Awlaki outpaced experienced al-Qaeda leaders by winning over jihadi digital audiences, which he did by delivering rapidly via cyberspace English-language zeal for the cause and justifications for violence to his supporters. Awlaki's online popularity reached such heights that leaders of al-Qaeda in the Arabian Peninsula considered stepping aside to allow Awlaki's rise to the top of the group. Bin Laden never warmed to the idea, and correspondence uncovered in the Abbottabad compound after bin Laden's killing showed al-Qaeda's leader reluctant to endorse Awlaki and critical of his rapid social media rise.

When Ayman al-Zawahiri recognized the fracture emerging in al-Qaeda's ranks after bin Laden's death, he immediately responded by trotting out veteran al-Qaeda ideologues offering expertise on jihadi Islam to support al-Qaeda's worldview over that of the emerging Islamic State. Abu Qatada, an alleged ghostwriter on one of jihad's first internet sites, spoke out in April 2014 as fissures between the two groups spilled into the open. He "denounced ISIS for 'ignoring instructions' from al-Qaeda emir Ayman al-Zawahiri," and further noted that ISIS fighters had been "misled to fight a war that is not holy."[3] Qatada's expertise represented nothing more than a blip in the jihadi social media ocean. I'd estimate that not a single ISIS member moved back into al-Qaeda's orbit based on any veteran cleric's offering theological insights. This same disrespect for Islamic expertise and arrogance among the new generation led my Twitter friend Omar Hammami to believe that after studying the Koran for only a handful of years and participating in the Somali jihad, he could craft an ideological vision for the future of all Muslims. His treatise was long, meandering, nonsensical, and yet a predictor of what was to come with the Islamic State's choose-your-own-jihad philosophies.

Instead of listening to Qatada and al-Qaeda's top leaders, Islamic State supporters drifted to their group's largely unknown emir, Abu Bakr al-Baghdadi. Will McCants, who, in addition to his work on the *Militant Ideology Atlas*, authored a biography of Baghdadi. In it, he explained that Baghdadi took a name bestowed upon him by his followers, " 'Commander of the Believers,' a title reserved for caliphs, the supreme spiritual and temporal rulers of the vast Muslim empire of the Middle Ages."[4] Before social media, declaring oneself a caliph might be blasphemy, but the Islamic State's virtual caliphate approved of Baghdadi's real-world claims. Baghdadi held some ideological credentials, but nothing compared with the traditional Islamic scholars al-Qaeda cited and coveted. Baghdadi and his followers furthered the justifications for his appointment to supreme jihadi leader by manufacturing a lineage tying him via bloodlines to the Prophet Muhammad himself. The more his followers said it, the more they believed it. The more they believed it, the more they'd say it.

The Islamic State's recruits didn't care about Zawahiri's thoughts on their future or listen to the clerics he deployed to scold them. Whereas al-Qaeda's recruits were often well educated and ideologically purists, ISIS boys were more criminal than pious, less educated and more zealous; committed to winning territory more than governing it right. For Western foreign fighters, much of their Koranic learning came from blogs, YouTube sermons, short social media bursts on Facebook, Twitter, or Telegram professed by passionate self-made experts who blended selected quotes from Muhammad with stylized presentation.

ISIS ideology justified any violence, garnering clicks and likes on social media. Enslaving women, beheading and murdering Shia Muslims in droves, or burning alive a captured Jordanian pilot could all be justified through the selective mixing of facts, scripture, and fantasy. Each of these acts would likely be rejected by the more ideologically pure bin Laden. At its peak, the Islamic State's young international foreign fighters served as de facto mayors and administrators of towns from central Syria to central Iraq. Their qualifications for the job were nothing more than their ability to enact violently toward populations they were not from and did not know or understand. The Islamic State's disdain for traditional expertise gave them the audacity to conquer, and the naïveté to try and govern.

The Islamic State's revision of jihadi doctrine came at the same time as the strongest anti-intellectual current in American history. President Trump's disdain for government elites and experts has its own hashtag: #TheSwamp. Most of his supporters heading into Election Day watched more television news than they read, and shared content with one another on Facebook and Twitter like never before. Their news sources were not the mainstream of educated elites. They substituted *Fox & Friends* for the *New York Times* and *Breitbart* for the *Washington Post*.

Trump's policy experts during the campaign were few, and those he selected were ones most Americans hadn't heard of and no other campaign would have selected. George Papadopoulos and Carter Page, while now well known in the context of the Russia investigation, were complete unknowns to the foreign policy community. Trump's policies and foci since assuming office have been at times deliberately against knowledge, research, and science. He has denied at times that Russia meddled in the U.S. election, despite the conclusions of his intelligence agencies, has withdrawn from international treaties and trade agreements, refutes climate change, called for investigations into voter fraud despite there being no evidence, and has even reignited conspiracies related to the connection between autism and vaccines.

While it may be customary to appoint U.S. ambassadors unfamiliar with the countries they're assigned to, Trump's appointees in many cases have no expertise at all in the departments and government positions for which they've been nominated. Trump's appointment to the Environmental Protection Agency, Scott Pruitt, doesn't believe in climate change. His initial nominee to cover the Department of Agriculture's science and research division, Sam Clovis, is an economist, not a scientist. (He withdrew because of his connection to the Russia investigation.) Ben Carson, a former presidential candidate and neurosurgeon, became the head of housing and urban development. Betsy DeVos, Trump's education secretary, has limited experience or credentials in education. The strangest of them all has been Trump's nomination of Brett Talley, a thirty-six-year-old lawyer, for a lifetime appointment as a federal judge in Alabama. Talley has never tried a case, is a relatively fresh graduate of law school, and is best known for writing horror novels and ghost stories. A year after Trump's election, the Associated Press analyzed his forty-three nominees in science-related positions and found that 60 percent held neither a master's degree nor a doctorate in a science or health field.[5]

For the ISIS boys, it was more important to have a caliphate than to do it right, more essential to pursue extreme violence than to effectively govern. For Trump supporters, it's more important to win than be correct, more important to be tough than compromise and move forward. And for America's liberal left, it's more important to be sensitive, all-inclusive, and politically correct, rather than pragmatic about partnering with those who reject some of their views. Their disdain for Trump supporters further reinforces the country's inability to reach compromise and achieve political progress.

Preference bubbles create a world of audiences where social media users, in many cases, identify with people they don't know over people they live with or see every day. We've seen the crowd not complement or compete with the core but seek to destroy it. Whoever gets the most likes is in charge; whoever gets the most shares is an expert. Preference bubbles, once they've destroyed the core, seek to use their preference to create a core more to their liking, specially selecting information, sources, and experts that support their preferred alternative reality rather than the real, physical world.

Moreover, social media preference bubbles increasingly shape our physical world. Our shared social media values lead to shared appearance, dress, purchases, living arrangements, and social norms. Preference bubbles have already started to shape real countries, from the Arab Spring to the Islamic State to the unraveling of the European Union with Brexit to the dismantling of American global leadership. And this—this is where Russia comes in. They see a dream come true in the West's migration into preference bubbles, an opportunity to use freedom of information, choice, and speech as wedges to turn cracks in Western populations into irreconcilable chasms.

During the Cold War, Russian intelligence articulated four steps for ideological subversion: demoralization, destabilization, insurgency, and normalization. Those phases never took hold in the United States during that period, but the methodology instead became part of the system of information management in the new Russia. Beginning in the late 1990s and extending through today, Russia implemented information overload on its population. Over time, the free press in Russia met with steep resistance from the Kremlin, up to and including imprisonment or death. In parallel, the Kremlin co-opted existing news outlets, intimidated others, and even created alternative news outlets that fawn over Putin's regime. When the active measures approach is applied on social media, this four-step process of ideological subversion moves at light speed. In the 1980s, Russian intelligence thought it would take active measures fifteen to twenty years—a generation—to achieve its effects. With social media, subversion now occurs in less than half that time.

Russia's method represents not an information war, but a war on information itself. The active measures goal of confusing fact and fiction is quite easy in the era of preference bubbles. To tumble democracy and subvert reality, the Kremlin can now—in just the same way it confused its domestic Russian audience—overwhelm American audiences with so much contradictory information that it becomes impossible to know what's true. Even if a social media user knows something to be true, an endless stream of contradictory explanations creates persistent doubt. Contradictory information bombardments inundate the audience, exhaust their comprehension, and lead them to withdraw from the information space confused, bewildered, and ultimately apathetic. Bewildered audiences fall back on their feelings, their biases, those they trust, those sympathetic ears and eyes in their preference bubble. The second stage of ideological subversion—destabilization—has already been achieved in America through social media.

The Kremlin doesn't even have to gather a lot of information on those they target; preference bubbles do it for them. Russia didn't create the Trump bubble or the Snowden groupies. Those bubbles created themselves and the Kremlin harnessed them. Each social media platform serves a purpose for active measures, and, through preference, Russia can help usher social media nations to information sources they've co-opted, repurposed, or even in some cases created to entice a useful audience. They use Twitter to infiltrate the preference bubble and reinforce useful narratives or spread new Kremlin ones. Facebook groups offer a circle of confirmation and implicit bias for saturating sympathetic audiences. Anonymous posting platforms like 4chan and Reddit offer the perfect platform for releasing *kompromat*, seeding ill-informed conspiracies suiting preference-bubble vulnerabilities, or rewriting history in support of false and alternative realities. LinkedIn is ideal for reconnaissance of foreign governments, defense contractors, and academia. Wikipedia is perfect for character assassination.

Through precise social media assaults and smear campaigns, Russian active measures assist in the death of expertise that might challenge their advances. Explanations for events, such as the shooting down of the Malaysian airliner over Ukraine, are challenged in every detail. *How do you know?* is always the first question. Anyone offering a truth must provide endless justifications. Sources of information, outlets, or witnesses offered as evidence are then badgered and challenged with follow-up questions. *How do they know?* is often the second subversive question, and this is shortly followed by a third challenge, *How can they be sure?* and finally, *Why do you trust them?* Each step of the way, question, question, question—just keep asking questions until the interrogation leaves the challenger battered.

After challenging the observations of Kremlin challengers, Russian trolls question their motives. *Who stands to gain?* they ask, suggesting that any act, statement, or explanation is designed only for self-serving purposes. There's no such thing as absolute truth, only different shades of public relations designed to benefit the purveyor of the information. Following degradation of the source, there will be an endless offering of alternative explanations; this is commonly known as "whataboutism." The trolls hound Kremlin adversaries with an infinite number of possibilities for any question, occurrence, or issue. The goal is to bombard the purveyor of truth with so many contradictory explanations that they must refute endless challenges to their information and provide evidence for why any and all challengers cannot be correct—an exhausting exercise leading many to surrender out of self-preservation.

Shifts in political support and the international order stemming from social media preference bubbles have been swift and volatile. Democratic movements birthed by Arab Spring protests have not excelled in Tunisia, have been overtaken in Egypt, largely failed in Libya, and have been defeated in Syria. The Islamic State's foreign mobilization, its creation of a caliphate, and its recent collapse spanned only four years. Republican support for Russia and Putin nearly tripled in only two or three years despite plummeting just a few years before with Edward Snowden's 2013 arrival in Moscow. Putin and his propagandists figured out the power of preference-bubble subversion before everyone else, and now bad actors worldwide have adapted the technique.

In the year and a half since President Trump's election, the world's authoritarians have rapidly adopted the Russian social media playbook for their own purposes. General Min Aung Hlaing powers Myanmar's military slaughter of the minority Rohingya population. Justification for his atrocities and popular support for his actions come in large part from the spreading of social media falsehoods. "No such thing as Rohingya . . . it is fake news," uttered an officer in Myanmar's Rakhine state security. The Rohingya are a long-persecuted Muslim minority that has been run out of Myanmar's western region. The *New York Times* at one point reported that roughly 90 percent of the information being spread about the Rohingya on Facebook was false.

The Philippines, where 97 percent of Filipinos have Facebook, may represent the most devastating example yet. Social media armies conducting "patriotic trolling" pushed President Rodrigo Duterte into office in 2016, and they have now been fully co-opted by his regime to attack his enemies and the free press. According to Lauren Etter of *Bloomberg Businessweek*, Facebook's stance on the Philippines problem "is simply that the country has come online fast and hasn't yet learned the emergent rules of the Internet."[6] For social engineers working with authoritarians, they go where opportunity arises, and there are dozens of countries like the Philippines that are ripe for dominating. Facebook now seeks to rapidly make policy adjustments to counter manipulative forces largely beyond its grasp.

Authoritarian regimes, political campaigns, large corporations—those with time, resources, access to social media activity, purchase information, and emerging technology capabilities—will infiltrate preference bubbles and deliver tailored messages, using tailored messengers, to move audiences to their preferred goals, agendas, history, and identity.

Social inception is the future. A hidden elite core will social-engineer an unwitting crowd into choosing the policies, politics, and preferences of the hidden elite, all without the crowd's ever realizing it. Future social media influence will be dominated by those who can aggregate and harness the collective preferences of audiences, deceiving the crowd into willingly selecting the core's agenda. If social media users believe that the core's agenda is their own preference, they'll wholeheartedly support an idea, a politician, or an organization. Social media nations and their members will cast doubt on experts who oppose the hidden core, insulate themselves from challengers through clickbait populism, and unwittingly support policies detrimental to their own well-being. In some ways this has already happened, with companies like Cambridge Analytica and the propaganda machine of Steve Bannon coming together during the election of 2016 to convince poor, working-class southern and midwestern whites to vote for a New York City real estate developer and reality TV star named Donald Trump.

The "hidden core" conducting social inception will win over key influencers by mapping their every purchase, chat, post, and picture, creating a targeting profile to nudge unwitting "useful idiots"—those enticed by money and ego—to advance scripted narratives. Media personalities, celebrities, and those aspiring to fame will be easily drawn to these social media inducements. The hidden core will enlist, empower, and advance fellow travelers—those enticed by revenge and ideology—as de facto agents for the cause. Political candidates, CEOs, or even clergy may wittingly join in with information power brokers. These agents will create the strong ties in the physical world—real human relationships—to unite the weak ties of the virtual world, creating a symbiotic preference bubble in pursuit of the hidden core's objectives.

Some of the social inception methodology I'm describing already occurs. It's what I wanted to do with Omar Hammami: nudge him to arrive at his own decision to denounce jihadism. My Twitter jousts with Hammami required reading spreadsheets of data, simple Microsoft Excel summations, and trial and error. Figuring out what worked and didn't work when messing with terrorists took many years. I had little time and only a few engagements to test out different conversational tactics. Today, I can still perform most of these social media actions from my house with a laptop and Microsoft Office. But I'd still be quite limited in my reach, and it would take a good deal of time to scale beyond single social media targets.

The social inception process will be relatively easy for well-resourced and motivated large-scale organizations around the world. They will be the ones who achieve data dominance, wield artificial intelligence, and rule social media's future. Data dominance occurs when an actor, either through corporate consolidation or third-party aggregation, can bring together sufficient social media preferences to clearly demarcate preference bubbles and individual motivations. Those sitting atop social media companies, mainstream media outlets, online retailers, and streaming services will have unimaginable power of persuasion as they gain more and more information on each person's daily life. Intelligence officers conducting espionage or targeting call this process a "pattern of life" assessment, but they've never had the capability now open to social media companies. The Internet of Things (IOT) records nearly every facet of one's life. Health applications tracking height, weight, heart rate, and steps, combined with Google searches and Amazon purchasing patterns, can provide social engineers with the ability to deliver someone a message via social media at the exact time and place where they are most vulnerable to it psychologically. A social media user won't know why they went for the content, but the big-data overlords will, which leads to the next emerging capability making this all a bit scarier: machine learning.

Artificial intelligence enthusiasts have always overestimated the point at which computers will overtake humans, become autonomous, and dominate our world. AI hasn't quite reached *Terminator* levels, but some of its tools have come into widespread adoption, and for social engineers, machine-learning advancements can be nuclear weapons for information warfare. *Machine learning* is the ability of systems to learn and improve from experience without being programmed by a computer operator. The information caches garnered through data dominance can be fed into machine-learning platforms, allowing the computer to learn more and more quickly—faster than any human propagandist ever could.

Machine learning applications will be able to pore over old data and current information feeds to design the perfect message for an entire preference bubble and the precise variant for each individual in the bubble, as well as when, how, and where to deliver it. Advertisers, political campaigns, and Russian disinformation peddlers would be handicapping themselves if they didn't use this approach to push their products and ideas.

Disinformation and misinformation have been easy to create and proliferate as the barriers to entry for these technological tools have lowered. False print stories and graphic memes have had a devastating effect. The latest dangerous technological dimension coming online now is falsified audio and video. Researchers at Stanford created software that allows programmers to spoof the speech and appearance of any person if given enough video and audio data. Presidents, elected officials, media personalities, and celebrities can be made to say anything a social engineer chooses. Data dominance will enable machine learning to create highly convincing hoaxes, propelling video smears and audio pronouncements that will move preference bubbles to alternative realities and false beliefs, playing to the benefit of those who design them.

I testified before Congress three times in 2017 and once more in 2018 regarding social media influence, and I immediately recognized from the experience that there are a few simple regulations that could impede these masters of manipulation. By law, political advertisements must announce themselves as such in print, radio, and television; these same rules should immediately be extended to social media, where political campaigns and super PACs increasingly focus their influence. But these regulations have gone nowhere, despite

Congress's grilling of social media companies in the wake of Russian meddling investigations. The rollback of net neutrality might also aid those with data dominance, as internet service providers could be coerced or influenced to shape content flows, targeting preference bubbles, denying users access, or restricting the content flows of those who challenge political propagandists. Emerging technology innovation and the regulatory environment, in sum, play to those able to dominate social media rather than users of social media.

Despite the ominous signs on the horizon, we must remember that people still have a choice. The backbone of preference bubbles, social media nations, and global commerce is ultimately trust. Social media influence could become more pronounced and yet less effective if users don't know what to believe. The educated and the mainstream may lose faith in social media, or maybe they already have, and in turn they could decide that the emotional crowds and social media mobs antagonizing each other in debates manufactured by influence peddlers aren't worth their time. Those not addicted to the constant stimulation of social media may simply leave the platforms altogether.

That said, there's also a more frightening possibility. The erosion of trust in social media may lead the less educated to rely more heavily on their preference bubbles, and the more educated to become apathetic and disengaged from democracy. If that happens, democracies will lose ground to authoritarians, science will retreat under attack from zealots, and countrymen will turn on one another, to propel their personal preferences over the greater good.

10

Surviving in a Social Media World

"We're in the business of filling souls," Howard Schultz told Scott Pelley during a 2006 interview on *60 Minutes*. Schultz, as head of Starbucks, saw a broader vision for the coffee company beyond specialized lattes. "We're in the business of human connection and humanity, creating communities in a third place between home and work."[1] The idea harked back to community gathering places of the 1950s, where people of all walks of life came together in the physical world and relaxed, talked, and even read books printed on paper. Starbucks innovatively offered specially selected music, which they repackaged, sold, and sometimes gave to customers as a gift while they waited in line.

Strangely, though, Starbucks's quest for community served as a harbinger of today's preference bubbles. Baristas created coffee cocktails to the exact specifications of each customer. In 2010, a spokeswoman for the company, Lisa Passe, explained to a *Wall Street Journal* blogger, "If you take all of our core beverages, multiply them by the modifiers and the customization options, you get more than 87,000 combinations."[2] Nearly everyone could have their own unique flavor tailored specifically to their liking. Tom Hanks's character Joe Fox, two decades ago, cleverly noted the challenge of so many choices in *You've Got Mail*:

"The whole purpose of places like Starbucks is for people with no decision-making ability whatsoever to make six decisions just to buy one cup of coffee. Short, tall, light, dark, caf, decaf, low-fat, non-fat, etc. So people who don't know what the hell they're doing or who on earth they are can, for only $2.95, get not just a cup of coffee but an absolutely defining sense of self."

As usual, Nora Ephron got it right. Preferences aren't about better taste, but identity. The movie also provided one of the first accountings of preference divergence and convergence between the virtual and physical worlds. The characters played by Meg Ryan and Tom Hanks meet and fall in love online, but they despise each other as corporate enemies in real life. The movie resolves with the two reconciling these differences—online preferences bridge real-world differences, ending in happily ever after.

This ending would be unlikely on modern online dating sites. Apps from Match to Tinder allow users to pick their preferences in the virtual world in order to streamline their matches in the physical world. Tom Hanks, the corporatist, and Meg Ryan, the small-business owner, despite living close by, would likely never see each other on today's social media dating apps. Their dating preference bubbles would ensure that the other never penetrated their social media fortress of personal preference. Even if they did appear in each other's feeds, the other's college, profession, or some additional Google research would likely lead one of them to "swipe left" on the other's profile. Ephron's brilliant script should instructively remind all of us, whether it's 1998 or 2018, that we don't really know what we want.

Fast-forward twenty years. Starbucks in the social media era has become the place where people come together to communicate with other people who aren't there. Lines to the counter fill with Americans donning headphones. It doesn't matter what music Starbucks plays; few even hear it. Heads are down, peering at smartphones; rarely do eyes meet. Customers might stand in the same line with the same people hundreds of times each year and never utter a word or even remember each other's faces.

The proliferation of smartphones and the rise of social media combine to allow any of us, at any time, to be in a world of our own preference, where we control the environment to block out anything we don't want or don't like. This only appears to be getting worse as more Americans work from home, conducting more conference calls than in-person meetings, talking more to Alexa, Amazon's in-home AI, than to real people and ordering food and groceries with smartphone apps rather than suffer the misery of shopping or dining out, surrounded by real humans. Self-imprisoned in our homes, governed by our preference bubbles, and yet seemingly having everything we ever wanted, Americans appear, at least on social media, to be absolutely miserable, downtrodden, victims of their excess and unfortunate to be alive. How do democracies, corporations, and citizens survive in this new and growing social media world?

\\\\\\\\\\\\\\\\\\\\\\\\\\\\

In 1835, Alexis de Tocqueville, the French diplomat, political scientist, and historian, published his seminal book *Democracy in America*. The work detailed Tocqueville's travels through the young United States in 1831, and his observations explained to the rest of the world why the experiment in American democracy might just be the best form of governance for balancing liberty and equality:

"Americans of all ages, all stations in life, and all types of disposition . . . are forever forming associations. There are not only commercial and industrial associations in which all take part, but others of a thousand different types—religious, moral, serious, futile, very general and very limited, immensely large and very minute."[3]

Tocqueville's study revealed much about the United States' charm, but most notably the power of cross-cutting associations between political and civil society that prevented government dominance over citizens. His observations about these horizontal linkages among citizens, rather than the highly vertical linkages of patronage found in autocracies and theocracies with high levels of inequality, demonstrated the advantages of America's melting pot.

Tocqueville explained what would later be described as social capital—how Americans come together in pursuit of their interests. Associations of marriage, religions, businesses, unions, military service, and schooling overlap to help protect each citizen's freedom and preserve opportunity for all, regardless of race, religion, socioeconomic status, or ethnicity. Robert Putnam, the renowned Harvard University professor who pioneered research into social capital, authored the famous and contentious book *Bowling Alone: The Collapse and Revival of American Community*. Putnam maintained that social capital was the glue for a thriving American democracy. By the late 1990s, though, membership in civic, social, and fraternal organizations was winnowing each year in America, and Putnam argued that the United States was headed in a dangerous direction. He pointed out that Americans were still bowling, but, rather than in organizations emblematic of civil society, they preferred to do it alone.

Putnam further defined the social capital concept pioneered by Tocqueville, dividing it into two types: bonding and bridging. Bonding capital arises when we Americans associate with people similar to ourselves. Bridging capital comes when we make friendships and associations with people unlike ourselves. Putnam argued that these two types of capital, when combined together, power American democracy, and that the latter's decline had already begun signaling an ominous future for the United States. Some challengers of Putnam's research, whose book was published in 2000, argued that he missed the new opportunities of connection and association offered by the internet. But these internet optimists of political science, similar to "long tail" enthusiasts, misunderstood how connections, when hyperpowered on social media, would strengthen Putnam's warnings of the degradation of social capital. Bonding capital has risen with the internet, but social media polarization has jarringly eroded bridging capital—the result being our current preference bubbles.

Post-publication, Putnam not only defended his thesis but worked to identify solutions for increasing American social capital, particularly of the bridging type. His 2001 Social Capital Community Benchmark Survey sought to discover approaches for increasing social capital but instead revealed more troubling indicators for American society. The study noted:

"Our survey results also make clear the serious challenges of building social capital in a large, ethnically diverse community. The more diverse a community in our study, the less likely its residents are: to trust other people . . . to connect with other people, even informally . . . to participate in politics . . . to connect across class lines."[4]

These findings come more than fifteen years before the social media nationalism arising today. It turns out that Tocqueville also saw this vulnerability in American society, as James Wood noted in a 2010 piece in the *New Yorker*:

In the book's second volume, he warns that modern democracy may be adept at inventing new forms of tyranny, because radical equality could lead to the materialism of an expanding bourgeoisie and to the selfishness of individualism (whereby we turn away from collective political activity toward the cultivation of our own gardens). In such conditions, we might become so enamored with "a relaxed love of present enjoyments" that we lose interest in the future and the future of our descendants, or in higher things, and meekly allow ourselves to be led in ignorance by a despotic force all the more powerful because it does not resemble one.[5]

Wood notes that Americans might not see the evil arising from a preference for the status quo, for confirmation of their beliefs, and for the implicit comfort of being among the majority. He continues with a warning directly from Tocqueville: "It does not break wills, but it softens them, bends them, and directs them; it rarely forces one to act, but it constantly opposes itself to one's acting; it does not destroy, it prevents things from being born."

Tocqueville wrote the definition of active measures more than 150 years before it was conceived. America touts its freedom and independence, but he noted, "I know no country in which, generally speaking, there is less independence of mind and true freedom of discussion than in America." He explained that "the majority has enclosed thought within a formidable fence. A writer is free inside that area, but woe to the man who goes beyond it, not that he stands in fear of an inquisition, but he must face all kinds of unpleasantness in every day persecution. A career in politics is closed to him for he has offended the only power that holds the keys."

Tocqueville concluded that the most educated and the most intelligent, those possibly best suited to lead the United States, had two choices: to toil away in intellectual circles to study the country's problems (i.e., academics), or move into the private sector and become wealthy (i.e., tycoons).

Democracy dies in preference bubbles. That's it, there's no way for Americans to communicate, debate, compromise, and thrive as these bubbles diverge and insulate themselves from challengers. The United States, if it stays on this trajectory, ultimately may not endure. I've explored social media preference bubbles in great detail, but they drive physical-world preference bubbles as well. We all increasingly live in places where we walk like, talk like, and look like one another. Members of the same social media preference bubbles move to places where they can reside with like-minded people who share the values, ethnicity, identity, and lifestyle of their social media nationalism. The Islamic State, while seen as an extreme in the West, provides an early example of this phenomenon. Social-media-induced fantasies led young Muslims, entire families of women and children, to voluntarily move to a war zone in Syria and Iraq—the digital tail wagged the physical dog.

Social media preference bubbles will continue to grow, expand, and merge with like-minded bubbles regionally and globally. Then they will harden by altering the real world to their liking. In November 2017, the *Economist* showed how Americans have become more politically polarized in the social media era. The number of those in the center of American politics, where policy compromise occurs, has dropped by nearly half since the dawn of social media.[6] It's not a coincidence that an American social media bubble advocates for a wall between the United States and Mexico—an actual physical bubble around the United States.

The liberal left cheers when Donald Trump blocks them on Twitter. They don't realize that his block isn't as much about protecting his ego, but about sealing off his base from any charismatic counter-arguments that might chip away at his popular support. Trump's Twitter challengers perhaps believe they are spotting and blocking Russian bots and trolls. In some cases they are, but in many cases they're simply blocking an aggressive member of Trump's base. The net effect on both sides is the same: increasingly divergent political poles, which makes policy compromise less likely and the future of democracy uncertain.

Traditionally in America, winners in elections may sow divisions on their way to victory, but after securing their position, officials move to unite Americans under a common banner, paving over divisions for the good of the country. This tradition has been abandoned in a preference-bubble world where each party and candidate remains incentivized, through gerrymandering and clickbait populism, to harden his or her voting bubble rather than bridge the divides in the electorate.

\\\\\\\\\\\\\\\\\\\\\\\\\\

Tocqueville noted that capitalism created many of the social connections, the civil society fabric, that bridged American communities.

Western businesses now suffer the consequences of negative influence campaigns. Dysfunctional democracies, which regulate these companies, sit idly by, uncertain how to help. My colleagues and I observed Russian active measures creating unease in foreign markets by advancing conspiracies. On one occasion, Russia's overt outlets and troll farms pushed a bogus terrorist attack conspiracy at Disneyland Paris, sending the company's stock price tumbling. Around Thanksgiving, the Twitter accounts tied to the Internet Research Agency tried to advance a conspiracy involving Walmart turkeys allegedly injected with poison. Russia's sporadic targeting of Western companies now provides the playbook for other nefarious social media influencers to target corporations.

Cyberattacks in support of influence campaigns will be particularly troubling for corporations. Hacks to secure *kompromat* will center around entities holding damaging secrets, namely banks and law firms, where transactions and confidential discussions take place. Corporate cybersecurity won't have to just protect against hackers seeking to steal money or intellectual property, but increasingly will need to protect their brand and trust with customers who can be lost to negative information campaigns powered by stolen communications. Insider threats in the form of disgruntled employees and foreign spies pilfering confidential documents, recording private conversations, and stealing potentially inflammatory emails will be of equal or greater concern to corporations than armies of hackers around the globe.

Social media companies should be in a complete panic. Trust in their platforms tumbled after the presidential election, and as more social media manipulators migrate to their platforms, governments will push to regulate them. The giants, particularly Facebook, YouTube, Twitter, and the ever-growing Instagram, may be destroyed by the nations, terrorists, and rapidly growing political manipulators frequenting their platforms for nefarious gain. Preference bubbles will continue to sour the average user experience and turn off those seek-

ing a civil discourse between friends and families. These will likely
be the same people with the money and resources to purchase the
higher-end products advertised on their platforms.

⁘⁘⁘⁘⁘⁘⁘⁘⁘⁘⁘⁘⁘⁘⁘

In 2008, when I was an expectant father, I sat at lunch with a friend
who was also awaiting the arrival of his first child. He told me that
his family planned to take a series of precautions to reduce the risk
that their child would have autism. They would insist on spreading
out the vaccinations their child received in his or her first year and
would verify that the shots were individually sealed, packaged one by
one rather than pulled from a bulk vat of serum. Individually drawn
and wrapped shots have fewer heavy metals, and if immunizations
were spread out over many months, a child would not at any point
receive large batches of injections that might tumble them down the
path toward autism, my colleague told me. He had spent a great deal
of time reading about this and had spoken to friends about it as well.

I did my own internet queries and decided to do the same—there
was a lot of research at that time suggesting that this really could
work. I planned my daughter's vaccination schedule and demanded
that doctors and nurses use pre-drawn, single-serving shots. The doctors and nurses had heard the hysteria from others as well. When I
launched into my concerns, they simply obliged. Pepper, my daughter, received immunizations in increments, all in the hope that she
wouldn't develop autism. I believed I was doing right by my child.

Pepper was a good-spirited baby who never slept more than a few
hours during the first year. She ate well and played on her own and,
aside from keeping her parents up all night, was a delightful child.
Pepper seemed to be on track by her first birthday, with no significant issues during medical checkups, at each of which, spaced apart
over many visits, she received shots. Pepper rolled when she was supposed to, crawled like other babies, and completed her first steps right

on schedule for a healthy kid. Except that, only a few days after she started walking, she began lifting her heel off the ground, tiptoeing across the kitchen and living room. I put socks on her feet, thinking the floor might be too cold. But socks didn't change her gait; she walked, just on her tiptoes.

Her funny first steps weren't the only concern. Pepper didn't respond to her name, either. At the one-year mark, I could shout her name in a room and her head wouldn't turn. Even if I clapped my hands as hard as possible right next to her ear, she wouldn't even flinch. Then, at other moments, when she heard running water from an appliance or when I drew her bathwater, she'd twist her ear to her shoulder, trying to block out the noise. I'd put Pepper down to bed and then frantically Google causes for what I was observing. She might be deaf, and her tiptoeing could be cerebral palsy. I was fraught with worry, and took her to the doctor for a hearing test. An echocardiogram showed her ears to be clear and fully functioning. The doctors said she should be able to hear just fine.

My Google searches had also pointed to autism. Pepper's symptoms—tiptoeing, delayed speech, unresponsiveness—were also included in the indicators for autism I'd read in articles. But I told myself that was less likely, because I'd taken the extra step of getting her shots spread out and taken precautions to keep her healthy. Each month, Pepper seemed more in her own world—happy, but tired and falling behind other kids developmentally. I didn't know why she was struggling, but doctors kept reiterating that it was too soon to jump to the worst-case scenario. "Some kids develop slower than others," they'd say, and Pepper, like many kids before her, might snap forward one day and catch up with normal kids her age. "This happens all the time."

The first gate for assessing children with autism comes at the eighteen-month mark. The exam is called M-CHAT, and it employs a multipoint assessment to determine whether children have devel-opmental difficulties. I packed the car for the short ride to Pepper's eighteen-month appointment, came back in, and gently put my hands on each side of her face. I looked into her big, beautiful blue eyes and said to Pepper, "All right. No autism, Pepper." With fingers crossed, we set off for the doctor's appointment, and a few minutes later we were in the middle of the exam.

Pepper failed the M-CHAT miserably. She was unresponsive and in her own world. Her cognitive abilities were extremely limited. Both her gross motor skills and her fine motor skills were poor. Despite her failing on nearly every indicator, the doctor seemed to want to kick the can down the road. "I don't know; she could still wake up tomorrow and start improving." But Pepper had done so terribly, and I stared in shock at the doctor, who seemed to be downplaying the assessment. I demanded a referral from him and, after some aggressive prodding, he reluctantly moved us on for another developmental assessment. About two months later, after we negotiated the complex web of medical referrals, Pepper arrived at Boston's Children's Hospital, where, fortunately, we encountered the world's best doctors in autism research. The team of doctors put Pepper through a rigorous exam. They held back their comments on exam day, but their faces couldn't hide what we all knew. A couple of weeks later, we returned to the hospital and received the diagnosis. Pepper had autism, and they recommended we start her developmental therapy right away.

"You need to give her vaccinations," the senior doctor, head of the autism group, told me as one of his pieces of guidance after Pepper was assessed with autism. "We've studied this extensively and, despite what you might hear, we've not found any link between vaccinations and autism." He looked me straight in the eyes and assured me that I'd be doing Pepper more harm than good by denying her the immunizations she needed against other diseases. Another doctor followed with more advice, noting that there were many claims about diets curing autism or some of its symptoms. The research teams were still exploring these diets and testing their merits, but at that time, their research team had not yet found any dietary treatment that could cure autism. "When we find anything that we think will improve or help Pepper, we will immediately let you know."

The vaccination link to autism was completely unfounded. My preemptive steps to decrease the chances of my daughter having autism did nothing to stop it. Nearly a decade later, I'm quite convinced that Pepper had autism the day she was born. I gave her the required vaccinations, but through a regimen of fear. Above all, I had fallen for "fake news."

Everyone falls for fake news sometimes, and if people say they don't, they either are lying to themselves, lack the humility to admit it, or still don't realize they've been duped. The vaccine-autism conspiracy continues on the internet and social media today. I've even seen it coming from Russian trolls infiltrating American audiences. Autism is something everyone hopes to avoid, regardless of political affiliation. If a troll can hook someone on an autism conspiracy, it'll be even easier to hook them on a social or political conspiracy. But why did I fall for fake news? I research social media and study this stuff, and yet I still convinced myself to space out my daughter's vaccinations and demand single packaged shots, just in case vaccinations caused autism.

I had created my own fake news back when I socially engineered prank calls at West Point as a cadet, and later I fell for fake news in trying to care for my daughter. I fell for the vaccine-autism conspiracy due to my preference bubble: my friend was the source, and I wanted to believe I could avoid autism by my own initiative. I suffered terribly from my implicit bias. Moreover, my friend and I both shared anxiety about increases in autism diagnoses and, as upcoming first-time parents, we both were more likely to spread the emotionally potent conspiracy.[7]

Social media has only made the danger of falling for fake news worse. Craig Silverman of *BuzzFeed News* provides superior analysis on social media's spread of false stories and rumors, and why I and everyone else can fall for fake news, particularly when we get emotional. His team analyzed a week of Facebook content during the lead-up to the 2016 presidential election, a polarizing, competitive, and emotional period. They compared right, left, and center media outlets reporting political stories that were being shared across communities and found two categories of content that did well: misleading or false stories and memes expressing partisan opinions.[8] Fake news outperformed real news, and visual expressions shareable on social media outpaced reporting—smaller bits of false content were seen more than longer pieces of factual content. The more people clicked on false information and shared false information, the more social media companies piped that type of content into their news feeds and those of their social media connections. Users' social media preferences and the desire of social media companies to fulfill those preferences created an entirely false reality. Even scarier, the artificial social reality selected by each social media user differs from any other. Each individual creates his or her own tailored perception of the world, with varying degrees of reality.

Social media can make a threat feel urgent in an unprecedented way. It instills in people a desperate need to access information through "breaking news," a term that heightens their fear and weakens their filter for falsehoods. Furthermore, the frequency of news sharing on social media further degrades a user's ability to block out falsehoods. That which one sees first and sees the most will be most believed. R. H. Knapp, who studied wartime rumors and propaganda, noted this phenomenon in 1944: "Once rumors are current, they have a way of carrying the public with them. . . . The more a rumor is told, the greater its plausibility."[9] Computational propaganda—the use of social bots by Russians and other political propagandists—provides a mechanism for broadcasting rapidly and repeatedly; the behavior that Knapp observed in the 1940s is exponentially more prevalent. Repetitive community sharing of confirmatory social media posts further distorts reality. Silverman notes, in research for Columbia

University's Tow Center for Digital Journalism: "With the emergence of digital social networks, our instant evaluation of a rumor can now be followed by a remarkably powerful act of push-button propagation. Once we decide that a rumor is worth propagating, we can do so immediately and to great effect."[10]

Social media applications and audience herding assist in the amplification of false rumors, but Nicholas DiFonzo and Prashant Bordia, two professors of rumor psychology, noted that well-crafted "rumors are believed to the extent that they (a) agree with recipients' attitudes (especially rumor-specific attitudes), (b) come from a credible source, (c) are heard or read several times, and (d) are not accompanied by a rebuttal."[11] The formula is simple for social engineers infiltrating preference bubbles: pick an audience, aggregate their preferences, create a story to suit audience preferences that matches an influence objective, add a source verifying a piece of the conspiracy, and then insert the content into social media and amplify it with social bots. If it's well crafted, the audience will "like" it, retweet it, and digitally share it with their social media tribe and even block out rebuttals to the false story.

The smartest propagandists and shrewdest ideological pragmatists, those engaging in social inception, will create their own news outlets and academic think tanks to manufacture the facts, research, and confirmatory science necessary to advance their agenda. Fake news is a fuel used by social media manipulators to power preference bubbles that will ultimately create serious dangers for all of society. Today's preference bubbles, which abhor compromise and debate, will use manufactured falsehoods not only to confirm their preferred beliefs but to stop the advancement of society as a whole. The alternative realities they create will slow down technological improvements, impede advances in medicine, and prevent the protection of our environment.

The fake news epidemic harms society as a whole, but social media itself may also be damaging the citizens addicted to it. Adam Alter's

2017 book *Irresistible* explains in great detail the dangers of internet addiction, and particularly social media addiction. The statistics are remarkable, and the implications obvious: "In 2008, adults spent an average of eighteen minutes on their phones per day; in 2015, they were spending two hours and forty-eight minutes per day. This shift to mobile devices is dangerous, because a device that travels with you is always a better vehicle for addiction." Alter then zeroes in on social media, noting, "Up to 59 percent of people say they're dependent on social media sites and that their reliance on these sites ultimately makes them unhappy." Why would we continue to use social media so much when it makes us feel so bad?

Social media, beyond preference bubbles, hurts users' critical thinking skills by shortening attention spans. "In 2000, Microsoft Canada reported that the average human had an attention span of twelve seconds; by 2013 that number had fallen to eight seconds."[12] Social media requires us to evaluate thousands of pieces of information each day, a massive exponential increase from previous analog generations, and now this is further compounded by attention deficits that severely damage our ability to successfully parse fact from fiction.

Addiction to devices leads to more consumption, but on top of that, it causes strange behavior outside the traditional social norms of in-person relationships. A University of Copenhagen study of more than a thousand Facebook users found that "the predominant uses of Facebook—that is, as a means to communicate and gain information about others, as habitual pastime—are affecting our well-being negatively on several dimensions."[13] The study documented how users suffer from "Facebook envy"—jealousy of one's Facebook friends, based on observations of their enjoyable experiences. A separate study, led by the University of Pittsburgh School of Medicine, noted an alarming rise in "Facebook depression," which occurs when "people who engage in a lot of social media use may feel they are not living up to the idealized portraits of life that other people tend to present in their profiles."[14] Dr. Brian Primack, who led the Pittsburgh study, added,

"This would be concerning, because it would imply that there is a potential vicious circle: people who become depressed may turn to social media for support, but their excessive engagement with it might only serve to exacerbate their depression." In social media, it seems, the more we learn about one another, the less we like one another.

The notable uptick in Facebook depression from the Pittsburgh study mirrored an online survey conducted by Kaspersky Lab. The results of more than sixteen thousand men and women from eighteen countries confirmed the findings of the Copenhagen and Pittsburgh studies and noted three additional conclusions ripe for social media influence. "People use social media as a forum for gaining social validation," Kaspersky noted. In addition, "people—men especially—become upset if they do not get the likes they hope for," and "in striving to receive more likes on social media, people can put themselves and their loved ones at risk by disclosing private information."[15] These survey results confirm what social engineers already know: play to a target's ego on social media and they'll be a useful idiot for the influencer and betray their friends and family for a virtual, temporary high.

I'm most concerned about the next generation, who now enter adolescence with a digital device strapped to their body—if they didn't already have one by early childhood. The next generation will have more virtual experiences than real-world ones. They will write more than they read, take so many photos that they will never look at again and possibly talk more to artificial intelligence than to other humans. The next generation will be able to recall any information with their fingertips, but they'll struggle in many ways to do anything of substance with what they access. Parents and elected leaders will have difficulty imparting wisdom that isn't contradicted by what's available on the internet and social media, some of which will be true and much of which will be false. It's quite possible that they will trust technology and artificial relationships more than real-world relationships.

Social media applications, in only a decade, have become an integral part of our life, offering many advantages. But they have created a destructive addiction and a collective anger that makes us, individually and as a society, worse off. How do we—democracies, corporations, and citizens—survive in this social media world?

\\\\\\\\\\\\\\\\\\\\\\\\\\\\\\\\\\\

I've offered some thoughts on how the U.S. government can protect Americans against Russian interference, but the threat to American democracy comes not from Russia but from America. The U.S. government will not save Americans from their preference bubbles, and since the election we've seen not just Russian active measures attempting to destroy our democracy but American active measures tearing down our institutions. It will take Americans fighting for their own democracy to fend off the social media manipulators, the hidden core, who seek to herd them and coalesce them into a movement outside of their control and only partly of their own design. Political and civil society must come together, leaders must emerge, and civil society must be rebuilt—on the ground, not online.

If we are to rebuild the civil society Tocqueville brilliantly observed in 1835, citizens must again form real relationships, physical encounters, by participating in neighborhood associations, providing public service, campaigning for office, attending in-person classes, and creating and supporting business associations. Compassion for and connection with those who don't look like you or talk like you comes from contact, not clicks or likes. Retired General Stanley McChrystal's recommendation of a national service, beyond the military, would be an excellent way to bring citizens together through common cause and shared values. Ultimately, real-world physical relationships will be the only way to defeat the online troll armies tearing democracies apart.

Social media companies sought to bring a great service to their users, and they did, but now they must help offset the unintended consequences of their products. Facebook, Twitter, and, to a lesser degree, other platforms have become the methods by which most Americans get their news, and much of it is false or manipulated truths. Facebook kicked off a new effort in 2018 to improve user experiences by reducing ads and news discussions in feeds and increasing content from friends and family. It remains to be seen whether "more cat pictures, less Trump talk" will bring audiences back to the platform, curb political animosity, and resolve differences. Democracies require a baseline of fact and fiction; otherwise, policy debate can't really occur and preference bubbles will continue to diverge into parallel fake-news-pumping factories. Google and Facebook initially sought to correct this problem by fact-checking news articles, but this approach quickly failed. Social media companies instead should work together to rate news outlets rather than fact-check each article.

My colleagues and I have offered for more than a year that the equivalent of *Consumer Reports* should be created for social media feeds.[16] Information Consumer Reports would be an independent, nongovernmental rating agency that evaluates news outlets across all types of media during a rating period. Outlets would receive marks based on their performance as assessed on two principal axes: fact versus fiction in the content it produces, and subjective opinion versus objective reporting. (The rating would then display as an icon alongside news links displayed in internet search engines and social media news feeds. The icon would provide the consumer with a virtual nutrition label for the outlet, one that, when clicked, takes the consumer to a web page where they can read for themselves why the news outlet received the rating it did as well as background on its funding, ownership, track record, retractions, and awards.

Such a rating system would not infringe on freedom of speech or the press, and it would rate both mainstream and alternative news sources. Those performing well, the ones focused most on reporting and facts, would likely garner more clicks, advertising, and subscriptions. Those assessed as poor would lose audience over time. Most important, this would put the responsibility on the consumer. If they read excessive fake news, they've no one to blame but themselves for being misled. Facebook, Google, and a handful of media outlets, in the year after the discoveries of fake news dominance, seemed primed to create such a system. But Facebook instead chose to put the evaluation to its users, asking the public to rate news outlets based on trust.[17] Having watched the public—the crowd—in recent years, I think it's equally likely that in this scenario, clickbait populism will reinforce rather than negate those outlets peddling less than factual information. The outlets with the most readers will likely rate those same outlets as most trustworthy, another version of tyranny by the majority, where confirmation bias can lead to alternative virtual worlds. Only time will tell whether the new Facebook system works, and, to Facebook's credit, at least it's trying something.

Social media companies can also slow the proliferation of disinformation and misinformation by ensuring the authenticity of accounts. Account anonymity may be of value at times, but social media companies must ensure that real people create accounts and can be traced as real sources of information. The costs to society of social bots far outweigh any benefits, and all social media platforms should end the creation and proliferation of automated, false personas. Twitter remains the most devastatingly negligent platform in this regard. It will have challenges making this policy adjustment, as its platform's design naturally comes with such vulnerabilities, and the company's value is tied to its volume of accounts. The killing of bots would put a huge strain on its finances. What if a celebrity, or an advertiser paying a celebrity to promote on Twitter, awoke to find that a third or more of their followers had disappeared overnight as a result of bot removals? Finally, social media companies design their applications to be addictive, but they must begin looking forward and ask whether their products have their users' best interests in mind. Increasing shares and advertising revenues makes sense only if the democracies and societies that allow these social media companies to operate remain standing.

Western democracies can also learn from one another—especially from those countries closer to the Kremlin that have built up resilience to propaganda. Finland fought Soviet disinformation for years, and Russian resurgence in this space led the Finns to develop a coordinated plan and trained personnel to deflect propaganda. They've also invested heavily in good public education, equipping their citizens not only to assess incoming information but also to recognize falsehoods because they understand how their own government institutions and processes work.[18] Americans enraged by WikiLeaks dumps, shouting claims of corruption or collusion, actually know little about the operation of the branches of government and the electoral process. Civics classes alone could enable Americans to better spot and ignore falsehoods, but in today's America, I imagine agreement on what should be taught during the class might be unachievable.

The Italian Ministry of Education, in preparation for its 2018 elections, joined forces with national broadcasters and digital companies "to train a generation of students steeped in social media how to recognize fake news and conspiracy theories online." Jason Horowitz of the *New York Times* described how students "will receive a list of what amounts to a new set of Ten Commandments. Among them: Thou shalt not share unverified news; thou shall ask for sources and evidence; thou shall remember that the internet and social networks can be manipulated."[19] Civil and political society can do much to ensure the integrity of institutions, restore a baseline of fact and fiction, and slow the divergence of preference bubbles, but ultimately it will be citizens, not governments, that save democracy and themselves from the new social media world.

Corporations have hurried to defend against hackers, and they must now extend those concerns to propagandists. Western companies should immediately develop response playbooks not just for cyberattacks but also for smear campaigns. Any company relying on customer trust in its brand will be particularly vulnerable to the spread of falsehoods via social media by nation-states, rival corporations, and protesters. When hacks occur, companies should assess not only what was lost, but how that information might be employed in a public smear campaign and how to inoculate themselves and their customers from the damage.

The quest for compromising materials to use against adversaries will make banks and law firms particularly hot targets. These sectors and their third-party data providers hold vast secrets for their clients that will be preferred pickings for those seeking to launch an information attack on an adversary. The Panama and Paradise Papers and the Sony breach are obvious examples of the dangers of maintaining potential *kompromat* on clients. Preparing against *kompromat* theft is twofold. Banks and law firms need to conduct more vigorous risk assessments of their clients to better understand how likely they are to be targeted in smear campaigns using private information. Taking on risky clients, ones targeted by foreign governments or interwoven into dark money networks, will likely make firms susceptible to increased hacker and insider-threat activity. This bleeds over to third-party data providers of firms operating in the darker worlds of politics and shady aspects of money laundering. Each company transferring data in and out of targeted hubs of *kompromat* is at risk of being an entry point for hackers seeking the secrets of the powerful.

Insider threats continue to grow as the greatest risk for data compromise and brand defamation. Monitoring internal threats and implementing controls should be a top priority as theft of secrets becomes the most damaging dimension of information warfare. Most companies have increased their technical monitoring, but, as I've discussed in this book, the human dimension—behavioral indicators—will be key in understanding when employees turn on their bosses and dump confidential information onto the internet.

\\\\\\\\\\\\\\\\\\\\\\\

Social media users can take several steps to survive in the modern social media world. First, and above all, ask whether the benefits of using social media outweigh the costs, and even if the answer to that question is yes, try to use social media less. Adam Alter proposed a solution for moderating social media and mobile phone use in *Irresistible*. The smartphone app Moment measures how much time a user spends sending emails, web browsing, using social media apps or making phone calls. I began using it and realized I average about the same amount of screen time each day as most users: three hours. My daily Moment update has helped me reduce my usage considerably in only a few months. Even before Moment, I'd undertaken social media blackouts at times, and the effects were almost immediate and positive. I feel as if I've got more time to accomplish tasks, my days feel longer and less frantic, and within a week I can generally concentrate more and feel significantly happier.

As social media users, we must all work individually to pop our preference bubbles. When we like something, we must always ask ourselves, *Do I like or retweet this image or news story because I believe in it and support it, or because I want to maintain my status in my social media crowd?* Clickbait populism can be a good thing collectively and individually, if users are doing it for the right reason, but as users, do we know why we are going along with the crowd? As Tocqueville noted, it's hard to challenge the majority, but it's necessary for the good of democratic society. When we click, like, and retweet, we must know why we are doing it.

Along those same lines, we must prevent the dangerous hardening of our bubbles. Trolls, those people who oppose our preferences, must be heard—don't block them. I let my trolls challenge me for a few reasons. Their challenges help me improve my arguments and sharpen my thinking. They provide me with needed reconnaissance for my enemies: the more I watch them, the more I can mess with them later. Also, trolls grow stronger outside my bubble than inside it—where *I can challenge them.* Above all, I like to keep my enemies close; that way, when I'm betrayed, I know who my opponents are inside and outside my digital tribe. I leave all my social media trolls in play, even when they make me angry. With one exception: those spouting racist, bigoted, and hateful speech that hurts my real friends and family must go.

Young people, in many ways, have figured out what their parents haven't regarding social media. Cornell University's Social Media Lab found that its students preferred Snapchat to its forerunners Facebook and Twitter because of the platform's "ephemerality"—their messages were temporary, not permanent, for all to see. Young college students can lower their inhibitions on Snapchat without worrying about their content being dumped in the open for public consumption. One student, Chang, noted, "The network is a lot smaller, and a lot closer. . . . On larger platforms, you have this large diverse group of friends. On Snapchat, it's very few and it's very selected. It would be your closest friends rather than a larger network of professors and other students that you met once in a class."[20] The younger generation's addiction might be worse, but they've learned from their parents about what can happen when you post something on Facebook or Twitter that you grow to regret. Maybe parents can learn from their kids and migrate to a more fulfilling platform like Snapchat, which brings together real friends and has design features that mitigate the spread of disinformation in ways Facebook and Twitter cannot.

We need experts in industry, government, and civil society to advance our nation. Everyone's an expert in something, just not in *everything.* We must respect and promote expertise, and each of us can do this individually by knowing when we're experts and when we're not, by separating our passions from our profession. We must also understand when our experts are suited to discuss a topic and when they are no longer in their particular domain.

We must be sure to pick experts who are good critical thinkers.

I've learned to evaluate them on three criteria. First, good critical thinkers have experience in their field and in the topics they discuss. Experience comes from years spent pursuing the discipline as well as certifications they've attained—doctorates, graduate degrees, and so on. Certifications have limitations, though. Not all are created equal—many highly certified experts have made dubious claims, and many who don't have credentials produce amazing insights. This leads to the second criterion: good critical thinkers have made many observations in their discipline or topic. They've conducted and published in-person research and study, served in organizations involved in fieldwork, and seen a variety of cases in the discipline for which they pontificate. Third, experts need to have gone through a deliberate process of analysis to arrive at their conclusions. They've taken a large number of samples, data, and observations and put that information into a structured evaluation approach, oftentimes assisted by technology and employing well-tested theories, to arrive at insights not easily discerned by a layperson.

I refer to this method of evaluating expertise—experience, observations, and analysis—by the acronym EOA. It helps me determine who to listen to and who not to listen to. When I hear pundits or even colleagues suggest that they should be listened to because they've "been doing this for twenty years," I usually stop listening, as they're not telling me how they arrived at their conclusions or where they personally gleaned their insights. If I'm in Washington, D.C., and I hear the same phrases and terms muttered at several different agencies by government officials pushing their expertise, I immediately ignore what they're saying, as they're playing to the crowd, pushing the clickbait populism of the Beltway.

On social media, the most effective way to challenge a troll comes from a method that's taught in intelligence analysis. To sharpen an analyst's skills and judgment, a supervisor or instructor will ask the subordinate two questions when he or she provides an assessment: "What do those who disagree with your assessment think, and why?"

The analyst must articulate a competing viewpoint. The second question is even more important: "Under what conditions, specifically, would your assessment be wrong?" This forces analysts to consider what they should be looking for so they can spot when their own assumptions are mistaken, when the alternative, rather than their assessment, is the correct one. If analysts haven't thought through their assessment, they won't be able to clearly indicate when they are wrong, and that's a strong signal that they've not explored all the possibilities or the merits of their challenger's argument. When I get a troll on Facebook, I'll inquire, "Under what circumstance would you admit you were wrong?" or "What evidence would convince you otherwise?" If they don't answer or can't articulate their answer, then I disregard them on that topic indefinitely.

When working in the U.S. government, I also learned how to evaluate information sources through a set methodology that provides a rubric for discerning what to listen to in a social media feed. I employ the acronym CMPP to quickly examine sources online or in person. The *C* stands for *competency*: Is the source of the information capable of knowing, gathering, or understanding the information they are providing? The *M* represents *motivation*: Why is the source providing the information? In the case of RT, Russia's English-language state-sponsored media outlet, the Kremlin's motivation is to advance its narratives and foreign policy among Western audiences while degrading its adversaries. The *New York Times* seeks to inform the public and sell newspapers. When the *Times* does its job well, these two objectives mutually reinforce each other.

The first *P* is *product*: What is the type of information being consumed? Print, audio, video, social media—each type of information can convey different meanings and impressions of reality. The most famous example of this effect came on September 26, 1960, when John F. Kennedy and Vice President Richard Nixon debated. Radio listeners thought Nixon had won the debate, but television viewers saw a sweaty, vacillating Nixon and instead thought the calmer and confident Kennedy had won.[21] The added dimension of emerging false audio and video will even further sway us emotionally in this fashion and has the potential to fundamentally misinform our understanding of the world.

The *process*, the final *P* in the acronym, asks how the information was acquired. Were the sources primary or secondary? Did the article or post represent its own research or simply pick and choose selected bits of data and opinions to support the overall conclusion? Did anything occur during collection of the information that could change or distort its meaning? Each of these process questions seeks to understand where biases and misinterpretations might be conveyed to the consumer.

Social media users must understand when crowds are smart and when they're dumb in order to know when clickbait populism moves society in a positive or negative direction. Crowds do well when they provide feedback on things they've actually experienced, when they are diverse in their perspectives and opinions, independent in their thoughts, and decentralized in their gathering of knowledge— everything today's preference bubbles are not. Clickbait populism, social media nationalism, and disregard for expertise make preference bubbles collectively dumb, particularly when they assess complex problems like war and peace in the Middle East, highly specialized disciplines like research into autism or climate change, and future-focused strategies and policies of which they have no EOA. When I see social media preference bubbles herding under these conditions, I look to the outliers, those individuals brave enough, as Tocqueville wrote, to challenge the tyranny of the majority.

We can each help ourselves by also understanding our enemies. In information warfare, don't fear David or Goliath, but Judas. Those you trust with your information, your secrets, those who once loved you or friended you on Facebook but are now jealous of your social media experiences or angered by your beliefs—they will be the ones to betray you. Know who you are sharing your information with and how they might ultimately use it against you. Even further, in the world of *kompromat*, where everyday information released under the guise of transparency or justice hits the internet and spins in social media, make sure to read the actual disclosure for yourself rather than let an influencer push it to you. If an article cites a study, read the study and make sure it describes the narrative of the news article before passing it on. If an allegation, such as an election hack, advances a narrative based on stolen information or leaked secrets, make sure you understand the context and have read the base information—stolen emails, confidential documents, whatever it might be—yourself before passing it on.

\\\\\\\\\\\\\\\\\\\\\\\\\\

The internet is not secure, and social media is public. I treat everything I type as if the whole world were watching, and increasingly I use phone calls or in-person meetings even when others want to do conference calls or chats. Encrypted applications like Slack, Signal, and Wickr have become my go-to communication methods. I've purchased subscriptions to the best newspapers and magazines. I've curbed my social media news consumption, further increased my cybersecurity measures, and prepared for the worst. And I accept that, despite all of these preparations, I'll never be fully safe on social media.

I wake up most mornings and wonder which of my enemies will strike back at me today. It might be a smear campaign from Russia, claims I'm a McCarthyite from transparency enthusiasts, charges of "deep state" from the alt-right, a knock on the door from a terrorist sympathizer pissed off about something I said on Twitter, a lawsuit threat from a political opposition group, or simply a betrayal from a Judas among my Facebook friends. Some of these things have already happened, more will follow, and each is not just likely but inevitable. Fame is fleeting, and infamy is forever, and on the social media battlefield, everyone is known for their biggest mistake, not their greatest achievement.

But I know my enemies, and I will mess with them. Social media sharks, hackers, and manipulators are all caught with bait. I'll pick some fights and lose them just so I can see who is attacking me. The penis picture they release of me someday, the one they hacked from my hard drive—that's not my penis. (I wish.) If they want to steal my personal information and tarnish my reputation, I'll confuse their facts and fictions about me. When they come to my information well, I put a little poison in it. When they infiltrate my Facebook and LinkedIn friends and follow me on Twitter, I let them in, and feed them some of what they want and some of what I want them to have. Those in my social media feeds should rightly wonder if they are part of a discussion, participants in an experiment, or trapped in an information battlefield where things are only partly what they seem. The answer to all of these questions is yes.

When you began reading this book, you may have had a different impression of me than you do now. Now, at the end, you may be confused, or even put off. Yes, I think like a terrorist, and that is why it's easy for me to talk to them. When I see my friends, I play practical jokes, and they in turn play them on me. Yelp is the app where I go to entertain myself with silly reviews, and when new social media applications come online, I imagine crimes that could be committed or people who could be duped by these new technologies. I often find more in common with terrorists on Twitter than I do with my neighbors, and I think more like a Russian propagandist than a Pentagon administrator. I'm a social engineer by nature, but I've had advantages and blessings that separate me from my enemies.

I grew up under the guidance of two great parents who instilled in me a strict sense of right and wrong from an early age. I attended a state-sponsored institution, the United States Military Academy, which imparted the values of my country, respect for the rule of law, and adherence to a code to protect and defend America's citizens from all enemies, foreign and domestic, and, surprisingly, the latter have outpaced the former in recent times. As with Assange and WikiLeaks, you should evaluate the measures I've taken based on their outcomes. Ultimately, I don't think you'll have much concern with my actions, as hopefully you now understand my intentions. I still believe in the American Dream, and I don't want my country or the rest of the world to be dominated by the privileged, the fortunate, and the corrupt who now use social media to suppress the very principles of freedom and democracy for which I, and many others, have fought and will continue to fight.

Finally, I know what I believe in, and why. I know there is good and bad in everyone, and that the direction they choose depends on the opportunities, resources, and freedoms they have to pursue the best version of themselves. American democracy, a highly imperfect system, remains the best chance for citizens to thrive and strive, and I will continue to support and pursue its best interests, in both the physical and virtual worlds. When you see me on social media, bantering, babbling, or battling, remember: I'm probably just messing with my enemies and pushing beyond my preference bubble. I hope you'll join me in doing the same.

Epilogue

Two men proudly sported "I'd rather be a Russian than a Democrat" shirts during a Trump rally in Ohio. The picture, retweeted thousands of times on Twitter and rocketing to the top of Reddit, demonstrated in a single image the monstrous success of Vladimir Putin's influence campaign on America.[1] The Kremlin's propagandists pitted Americans against one another in unimaginable ways prior to the advent of social media. As discussed earlier in the book, the goal of any influence campaign is to create a behavior change in the targeted audience. Today, two years into the Trump administration, some Americans openly advocate for Russia over their fellow Americans, and advance Russian foreign policy agendas to the direct detriment of decades-long alliances and partnerships, many of which have aided and continue to aid the United States in its wars in Afghanistan, Iraq, and Syria. Elevation of President Trump and the resulting launch of the Special Counsel investigation into collusion and obstruction have further ruptured the divide in the United States. Persistent political infighting and retreat from world leadership have made the world a playground for authoritarians. Russia, China, Iran, Philippines, Saudi Arabia, Myanmar, Cambodia, and likely many more countries have been outed for nefarious influence employing social media to either suppress opposition domestically or move audiences abroad. The list will continue to grow.

It's been exactly one year and two days since I completed the first draft of *Messing with the Enemy,* and we all know a frightening amount more about the bad things happening on social media. It's hard to determine where to begin, but I'll start with Vladimir Putin and his propagandists.

In February 2018, Special Counsel Mueller indicted the Internet Research Agency, LLC, in the most revelatory filing to date. The indictment described in remarkable detail how Russia created its social media cutout to sow chaos in the American electorate.[2] In total, thirteen Russian nationals and three Russian entities were charged. Accusations included conspiracy to defraud the United States and conspiracy to commit wire fraud, bank fraud, and identity theft. The indictment described how Russians traveled to the United States to set up technical infrastructure, how they lured Americans to protests on both sides of an Islam rally in Texas, and how they staged an event outside a Cheesecake Factory in Florida. The evidence proved nearly impossible to challenge for those previously in denial about election interference. Investigators revealed intercepted Internet Research Agency emails describing when they knew they'd been caught. Irina Kaverzina, a troll farm member, emailed around September 13, 2017: "We've had a slight crisis here at work," she wrote. "The FBI busted our activity (not a joke)."[3]

Five months later, Special Counsel Mueller's team dropped the next logical installment, revealing the hackers behind Putin's influence machine. On July 13, 2018, an indictment revealed the names of Russia's hackers embedded in their Main Intelligence Directorate of the General Staff (GRU).[4] The indictment clearly laid blame for the theft of Clinton Campaign, DCCC, and DNC documents and emails on the Russian government. Allegations confirmed what had been assumed by many for years, that Russia provided stolen information, "maintained by an organization ('Organization 1'), that had previously posted documents stolen from U.S. persons, entities, and the U.S. government."[5] All analysis points to Wikileaks as "Organization 1." The document further outlined the structure, staffing, and function of different parts of the GRU's cyberattack teams and offered the outside world, in only twenty-nine pages, great detail on what had previously been a mystery in the cybersecurity world.

Not spelled out in the indictment but swirling in the open source were claims that it was maybe not the Americans but an unexpected European ally that cracked into the GRU. Dutch newspapers reported a few months earlier that the Netherlands AIVD intelligence service tapped into one of Russia's hacking collectives, Cozy Bear, as early as 2014. Over several years, the AIVD hackers monitored Russia's hackers. The Dutch account alleges the AIVD watched cyberattacks on U.S. targets in real time and notified American intelligence agencies of the breaches, offering evidence of Russian attribution. An amazing example of how important America's allies are to American national security at a time when President Trump chastises NATO and the EU and curries favor with President Putin.[6]

The breadth of the U.S. investigation into Russian social media interference continued to grow by the fall of 2018. Prosecutors in the Eastern District of Virginia filed a criminal complaint against Russian Elena Khusyaynova, the accountant for the Internet Research Agency (IRA).[7] The complaint outlined the corporate cutouts employed by the Russian troll farm to conduct their social media influence and noted their nefarious influence didn't stop with the 2016 presidential election. Continuing into 2017, the IRA inflamed American divisions on issues from "immigration, gun control and the Second Amendment, the Confederate flag, race relations, LGBT issues, the Women's March, and the NFL national anthem debate." The filing illustrated the Kremlin's efforts were cost effective but not exactly cheap, with expenses between January and June 2018 totaling more than $10 million.[8]

The final installment in troll farm revelations surfaced at the end of 2018. The Senate Select Intelligence Committee released two independent reports that analyzed tens of thousands of Facebook, Instagram, and Twitter posts and ads known to be attributed to the IRA. The reports' findings mirrored the findings of my social media team's work since 2014. Russia used nearly every social media platform to infiltrate, divide, and inflame the U.S. populace along racial, religious, social, economic, and political lines. Russia's trolls peaked their activity at important times in U.S. politics and international events.[9]

Insights into Putin's online offensive have been informative and revelatory and have confirmed what Weisburd, Berger, and I observed starting in 2014. But more important and fear inducing has been the Special Counsel's revelations about the connections between the Kremlin and the Trump campaign. By the close of 2018, five ex–Trump aides had struck plea deals with the Mueller team, and the *Washington Post* had identified at least fourteen Trump associates interacting with Russians during the campaign and the transition to the White House.[10] Nearly weekly revelations have shown how Russian agents, oligarchs, and former and current Kremlin officials connected with and pushed along the Trump campaign. President Trump's repeated denials of business dealings with Russia, made during numerous public appearances and tweets, appeared to be false when his former lawyer Michael Cohen conceded in his guilty plea to a U.S. District Court that he pursued a Trump Tower Moscow project through June 2016.[11] Whether it was online or in person, Russia's campaign to elevate Trump and eliminate Clinton was intense, determined, and continues to roil America at home and abroad. President Trump's placating of Russia, assuaging of Putin to the detriment of America, and erratic and bizarre response to the Special Counsel Mueller investigation have only furthered the success of the Kremlin's influence operation.

The U.S. response to Russia's social media influence campaign has been shockingly slow. Never has any attack on America and its democracy been discussed so much and generated so little action. President Trump's denials of Russian interference challenged the ability of U.S. institutions to mount adequate protections for the 2018 midterm elections, and likely further emboldened Russia to continue its manipulation. Luckily, committed public servants in the FBI, the Department of Justice, and the National Security Agency and increased efforts by the social media companies thwarted Russia's limited attempts in 2018.

Russia again employed its three-step cyber approach of hacking, infiltration, and influence beginning in 2017 but only on a narrow set of targets. In August 2017, staffers in Missouri senator Claire McCaskill's Washington office received strange emails.[12] The messages suggested users' Microsoft Exchange passwords had expired and that they needed to create new ones. If they had clicked on the link, these staffers would have been sent to a fake log-in page for the Senate's Active Directory Federation Services, requesting they reset their password. The spearphishing campaign was eerily similar to the one that nabbed the emails of Hillary Clinton's campaign manager in the run-up to the 2016 election. McCaskill, an opponent of Trump who routinely drew the president's ire, did lose her seat to a Republican in Missouri but her team didn't fall for the hacking attempts.

McCaskill wasn't the only Democrat fighting off what appeared to be Russian incursions. Senator Jeanne Shaheen of New Hampshire received a similar spearphishing attempt that fall and a curious phone call.[13] Someone impersonating a Latvian official tried to set up a meeting with the senator to discuss Russian sanctions and Ukraine. Her staff smartly called the Latvian embassy and learned through a physical verification that no such person existed. Alongside the McCaskill and Shaheen attempts, Democratic candidates running against Dana Rohrbacher, the Republican congressman friendliest to Russia on Capitol Hill, received hacking attempts as well. But in 2017 and 2018, America's cybersecurity and overall awareness was much better, and Russia's attempts were unsuccessful.

The social media companies were better by 2018 too. Facebook placed the greatest resources on countering disinformation, and their efforts bore some fruit. During the summer leading up to the election, Facebook connected a Russian Internet Research Agency Facebook account to a separate, newer set of inauthentic activity, leading to the shutdown of thirty-two false Facebook and Instagram pages and profiles suspected of a Kremlin connection. Facebook didn't explicitly name Russia as the culprit of the nefarious social media activity, but the creation of fake pages such as "Black Elevation" and "Resistors" appeared to be infiltrating minority groups in opposition to the Trump administration, encouraging protests, and even posting a job ad for an event coordinator.[14]

Even if Facebook's efforts weren't working, it is harder for a foreign government to influence the American electorate in state and local races than it is in a presidential election, when Kremlin themes and messages can more easily be shared across national platforms. In the final days leading up to the 2018 elections, Russia's troll farm, unsuccessful at influencing the elections, took a different approach from two years before. They created numerous Facebook, Twitter, and Instagram pages claiming to be backing both popular Democrat and Republican candidates. Facebook conducted a last-minute takedown of the accounts. Just before election night, a website entitled "Internet Research Agency American Division" popped up. The site taunted U.S. intelligence agencies and the public, listing one hundred Instagram and Facebook accounts they claimed to control, stating, "These accounts work 24 hours a day, seven days a week to discredit anti-Russian candidates and support politicians useful for us than for you."[15] The entire effort likely represented a false flag and followed the general rule of Russian propaganda. When the Kremlin is being secretive and stealthy, they are trying to hide their hand in something they are doing, i.e., 2016. When the Kremlin is being loud and sloppy, they are trying to convince the world of something they are not actually doing, i.e., 2018.

Americans, whether it's the last war or the last campaign, tend to think the next contest will look like the last. But there's only very scattered evidence that Putin's masters of manipulation are pursuing the same tactics they used in 2016. The state-sponsored propaganda continues, social bots proliferate, but their effectiveness has lessened and Americans have grown wary of not only political messaging but social media itself. The Kremlin has to evolve to win and sustain American audiences, and their influence has moved rapidly from the covert to the overt.

Former CIA station chief in Moscow John Sipher summarized the approach best during the summer of 2018 in his article "Convergence Is Worse Than Collusion."[16] Sipher noted, "Convergence can be defined as distinct groups doing the same things for different reasons, or as a unity of interested evolving from separate starting points. Both Putin and Trump seek to inject chaos into the U.S. political system. They support an assault on U.S. foreign policy elites, encourage fringe and radical groups, and envision a United Stated untethered from traditional allies. They also share a willingness to utilize informal and semi-legal means to achieve their goals."

The cascading effect of Trump and Putin convergence reverberates daily in American discourse. Whether a "Useful Idiot" or a "Fellow Traveler," President Trump acts as an unprecedented agent of influence for Russia in America. Mysteriously, on three separate occasions President Trump has repeated Russian disinformation on highly specific foreign policy issues relevant to the Kremlin.[17] Shortly after his inauguration he parroted Kremlin claims of Polish aggression toward Belarus. He later shoved the Montenegrin prime minister at a NATO summit after claiming on Fox News that Montenegro is an aggressive country and asserting Americans should not go to war on their behalf. Finally, in January 2019, President Trump repeated revisionist Kremlin propaganda as to why the former Soviet Union went to war in Afghanistan. In all three cases, President Trump appeared to reference not his intelligence community, established facts, or mainstream analysis but the disinformation of Russia. Where the president gets this information is unknown and absolutely concerning for America's national security and its allies.

Russia's overt influence campaign continues to plot and push against the U.S. populace, its political parties, social groups, and politicians. Russia's official state engagement is traditional and direct;

meetings, summits, and exchanges follow traditional state-to-state protocols. The most devastating event was undoubtedly the Helsinki summit in July 2018. President Trump stood onstage alongside the Russian president and accepted Putin's denials of election interference while refuting the U.S. intelligence community. The event sent shockwaves throughout the world and caused alarm back in America, fueling arguments that President Trump might be compromised, coerced, or owned by Putin.

This foreign policy disaster did not occur in isolation. Just weeks before, diplomatic engagement and congressional exchanges with Russia restarted after many years of dormancy. President Trump's disastrous Helsinki summit with Putin left American audiences aghast as the commander in chief took the denials of Putin over the assertions of his U.S. intelligence community. But this didn't slow Senator Rand Paul, who visited Russia delivering a follow-up letter from Trump to Putin calling for the two countries to work together on "countering terrorism, enhancing legislative dialogue and resuming cultural exchanges."[18] Exactly the types of connections the Kremlin hoped to build with America by elevating Trump.

These traditional diplomatic engagements are neither nefarious nor the Kremlin's main effort to weaken America. Where the Kremlin has really innovated is in its attempts to win over state and local parties within the United States, like the National Rifle Association and the Christian community—and, crucially, individuals who might, for one reason or another, be sympathetic to Russia. Business leaders, up-and-coming politicians, and media figures all provide authentic American vessels for advancing the Kremlin's worldview in the U.S.A. This is the real goal the Kremlin pursues today, and its seeds were planted in plain sight just a few years ago.

For some years, the Kremlin has chosen certain individuals, peripherally aligned with its state security complex, to spread their messages in the United States. One prominent example is Maria Butina, who, at twenty-three, founded an organization called the Right to

Bear Arms. From a small underground gun range in Moscow, Russia, she seemingly invented a gun rights movement in a country with no constitutional right to bear arms.[19] By 2014, Butina appeared as a young Russian gun rights activist in America, surfacing at the national NRA convention, visiting colleges, delivering speeches, and networking with prominent members of the GOP. In 2017, she attended a post-Thanksgiving barbeque at the home of South Carolina Republican lawmaker Mark Sanford, and throughout that time, she was going to Grover Norquist's weekly gatherings in Washington, D.C.[20] When the FBI searched her apartment in April 2018, they found evidence suggesting she'd been corresponding with agents in Russia's intelligence agency, the Federal Security Service (FSB).[21]

Butina and her boss, Aleksandr Torshin, a former Russian senator and deputy governor of Russia's central bank, had also been working to infiltrate America's Christian community. Their chosen tool was the Russian Orthodox Church, a natural ally for the American evangelical movement: both have strong conservative views on gay rights and abortion. Fostering these kinds of connections could, the reasoning went, help the Kremlin shape America's view of Russia. In 2017, Torshin and Butina attended the National Prayer Breakfast.[22] Prior to attending, she recommended to a colleague that President Putin might attend the gathering and meet President Trump. Two years later, an estimated sixty representatives from Russia's religious and political elite, CNN reported, attended the National Prayer Breakfast—along with President Trump.[23]

White nationalists were another group targeted by the Kremlin to foster sympathy for Russian views. David Duke, a onetime leader of the Ku Klux Klan, lived in Russia for five years and claims he sold his book in the Russian parliament, the Duma. He was also invited to attend a white supremacist conference in Russia, bringing many American white supremacists to Moscow. These "cultural exchanges" did what the Kremlin must have hoped they would: Duke and his allies began praising the country, calling Russia the "key to white survival"[24] and, during an alt right rally in Charlottesville, Virginia, in October 2017, chanting "Russia is our friend."[25] Based in Alabama, the white supremacist organization League of the South now offers a "Russian Outreach" section on its website.[26] These Russian language pages include summaries of the group's basic beliefs, a page dedicated to conspiracies of restricted European immigration to the U.S., and the potential for alliances between the American South and Russia against globalists.

The Kremlin also runs official cultural exchanges and diplomatic programs. In August 2018, Russia's Foreign Ministry appointed American actor Steven Seagal, whose martial arts movies are popular in Russia, to "facilitate relations between Russia and the United States in the humanitarian field, including cooperation in culture, arts, and public and youth exchanges."[27] Just a few weeks earlier, after years of dormancy, it also relaunched congressional exchanges with Russia. In July, eight Republicans visited Moscow—the first congressional delegation to visit the country since Russia annexed Crimea in 2014. Legislative exchanges with the U.S. and other countries are not unusual per se, but reigniting relations with the Kremlin amid two congressional investigations and a Special Counsel investigation into Russian election interference seemed strangely timed. The GOP statements regarding Russia only confirmed the exchange as a bizarre spectacle. "Most countries would meddle and play in our domestic elections if they could," Republican Senator Richard Shelby of Alabama told the Washington Examiner in June 2018, just before he'd leave for Moscow. "We've done a lot of things too."[28]

To build goodwill with these Republican officeholders, Russia has developed a simple playbook: attack Democrats and repeat Republican narratives. In March 2018, Russian aluminum magnate Oleg Deripaska, for example, published an op-ed on the conservative U.S. website *The Daily Caller*.[29] Deripaska took aim at President Obama's assistant secretary of state Victoria Nuland, Democratic senator Sheldon Whitehouse, and former DNC committee chairwoman Donna

Brazile and echoed the Republican conspiracies regarding Fusion GPS, billionaire George Soros, the "Deep State," and the investigation into Russian interference in the U.S. presidential election of 2016.

If there's more of this—of American gun rights groups; Christian religious summits; white nationalists; or even motorcycle enthusiasts surfacing at events in Russia, or vice versa, with Russians surfacing at events in America—we'll know that the Kremlin plan is on pace and on target. Putin won't have to topple the NATO military alliance, break the European Union, force the U.S. from the world stage, or instigate paralyzing political conflict. Instead, the Kremlin's newly won allies in America will continue to willingly advance Putin's subversive quest for global power.

As disastrous as the Russia influence efforts have been on the United States, they may have been even more devastating to the social media companies. Immediately following the presidential election, the social media companies suggested there was essentially no Russian manipulation on their platforms. Time, investigation, and research have shown again and again that social media platforms were rife with disinformation.

Senator Mark Warner, the top Democrat on the Senate Select Intelligence Committee investigating Russian interference, described Twitter's initial presentation in September 2017 as "inadequate on almost every level."[30] Twitter had shut down only 201 accounts linked to Russian operatives on Facebook. They seemed unsure of what to even look for. But within a year, Twitter seemed to catch up, and on October 17, 2018, they released the full data sets of 3,841 accounts affiliated with the Internet Research Agency and 770 accounts potentially originating in Iran.[31] The release immediately became the focus of dozens of disinformation study groups springing up in the wake of the 2016 election. More interesting has been Twitter's call for proposals in March 2018 to measure the health of public conversation.[32] Twitter's CEO Jack Dorsey routinely asserts the social media platform is a virtual town square, but the company seemed to realize their forum had turned into an inhospitable preference bubble war zone.

YouTube provided an essential gateway for RT (a Russian government–funded international television network) to broadcast its foreign language propaganda to Western audiences. YouTube, the world's largest video hosting platform and second largest search engine, received scorn for being the world's home for conspiracy theories, misinformation, and state-sponsored propaganda. In 2018, YouTube remained a prominent venue for RT to spread disinformation regarding the Syrian war and the GRU's alleged poisoning in the United Kingdom of former Russian military intelligence officer Sergei Skripal.[33]

Google as the parent of YouTube suffered criticism from the U.S. political right for alleged bias in its search engine. Google didn't attend the September 2018 Senate Intelligence Committee hearing alongside Twitter and Facebook, drawing the ire of congressmen and the public. Later in December, in a Senate House Judiciary hearing, congressmen didn't seem to fully understand how search engines and Google as a company work.

Facebook took the greatest leaps to tackle its fake news problem, focusing on three key areas to improve the integrity of information on its platform: disrupting economic incentives, building new products to curb the spread of false news, and helping people make more informed decisions when they encounter false news.[34] The company routinely closed accounts in 2018, tested and innovated new fact-checking methods, worked with governments to counter inauthentic activity, and constructed an elaborate war room for the midterm elections. Even when dragged in front of Congress, Facebook founder Mark Zuckerberg, known for public awkwardness, sailed through testimony, fully overmatching his congressional interrogators, who seemed ill equipped or laughably out of touch with regard to how today's technology works. Facebook's second-in-command, Sheryl Sandberg, did equally well, and it looked for most of the past year as if the public's and regulators' ire toward Facebook would wane. That was until mid-November 2018.

The *New York Times* dropped a bombshell on Facebook's comeback, asserting that Facebook, facing user backlash, employed an opposition-research firm to discredit detractors by linking them to George Soros, a wealthy liberal donor who's been the target of personal attacks and fake news on sites such as Facebook.[35] Hiring of the public relations firm sparked public outrage undoing much of the progress Facebook had made countering disinformation since 2016.

Two years of public outrage and congressional hearings would lead one to believe that government regulation would have been instituted, but that would not be true. Every congressional session with the social media giants showed just how unlikely lawmakers are to effectively regulate tech companies. Mark Zuckerberg, Sheryl Sandberg, Jack Dorsey, and Google CEO Sundar Pichai—introverted techies—all handled congressional grilling well and outmatched the legislators who would regulate them. Even when bipartisan support for legislation exists, Congress has been unable to move laws forward to help defend the 2018 midterm elections. They failed to pass the Honest Ads Act, which would "prevent foreign actors from influencing our elections by ensuring that political ads sold online are covered by the same rules as ads sold on TV, radio, and satellite."[36] Even more surprising, the Secure Elections Act did not move forward. GOP sponsor of the bill Senator Lankford said the measure "helps the states to prepare our election infrastructure for the possibility of interference from not just Russia, but possibly another adversary like Iran or North Korea or a hacktivist group."[37] I can think of no time in American history when public and government attention and outrage have been so high without seeing any real progress toward solving the problem. Congress is unlikely to stem the tide against disinformation, and American elected officials have become the source of as much misinformation as authoritarians.

Facebook's public relations counteroffensive mirrors what I expected as I finished writing this book. Moving forward, the worst abusers of social media for nefarious influence will be public relations firms and political campaigns. The midterm elections of 2018 showed how Russia's art would be quickly adopted by the most aggressive and best resourced. Russia didn't push much disinformation to influence the 2018 elections, but even if they had, it would have been completely outpaced by the endless volumes of false information peddled by political campaigns and the White House. President Trump and his staff took to both the mainstream and social media to advance bogus conspiracies about a human refugee caravan invading the U.S. southern border leading up to election day 2018. Despite being thousands of miles away, President Trump shared unrelated or misleading footage of violent migrants completely detached from reality. In a separate instance, President Trump ejected CNN reporter Jim Acosta from a White House press briefing. Sarah Sanders, White House press secretary, then shared a doctored video on Twitter seeking to justify Acosta's dismissal for aggressive action. The video had been sped up and the reporter's comments had been removed.[38] The rapid proliferation of social media disinformation was expected, but rapid proliferation of social media disinformation from the White House was not. If America can't count on the commander in chief to do the right thing, we most certainly can't expect everyone else to do much better.

The best disinformation peddlers in the future will have three distinct technological advantages over those that came before them. Cambridge Analytica demonstrated how the aggregation of user data provided deep insights for nimbly targeting and influencing selected audience members. Future firms and campaigns able to employ machine learning to mine and connect user data across nearly all aspects of a person's daily life. Knowing individualized, intimate details will enable the best tech-enabled manipulators to subtly nudge social media users in undetectable ways.

Social bots continue to improve and will be even more lethal with artificial intelligence. They'll not only be able to communicate with people and be indistinguishable from real humans, but they'll be able to communicate with one another in ways that are indistinguishable from real crowds. Bot communities will further disorient users' perception of reality.[39]

Finally, the White House tweeting of altered video will quickly be outpaced by sophisticated fake audio and video content. Bogus multimedia content, known as Deep Fakes, has dramatically increased in volume, availability, and sophistication just since I began writing this book and threatens not only democratic governance but also public safety. The most poignant example of the deadly implications of rapid fake news proliferation has occurred in India, where WhatsApp groups incite mob killings.[40]

The year 2018 offered a glimpse of where our social media preference bubbles plot a course for further division. American trust in social media platforms declined precipitously. A Pew research study released in September 2018 found that 42 percent of Facebook users have taken a break from the site in the past year. Forty-four percent of Facebook users between eighteen and twenty-nine years of age deleted the app from their phone in the past year, more than triple the rate of users over sixty-five years of age.[41] Trust in social media–derived information declined across all demographics and segments during the year. This might represent a positive trend if democratic citizens return to physical relationships and engagement. But this remains difficult to measure, as disengagement on social media may just represent disengagement in general, leading to apathy toward democracy. Public, political activism and voter turnout in the 2018 midterms appeared higher than ever, though. Maybe Americans are turning a corner, both online and in person.

A two-pronged danger sits on the technological horizon, though. Younger generations spend increasing amounts of time in the virtual world, not just in social media, but also substantial amounts of time playing video games. Their allegiances to virtual connections and emerging artificial intelligence may be greater than to friends, family, and fellow citizens in the real world. How can democracies, their leaders and their institutions, connect, engage, and inspire future generations when they aren't participating in the physical world? Alongside the separation between real and virtual is the balkanization of the internet. The endless pursuit of preference bubbles has sent audiences into increasingly tailored and fragmented spaces suiting their desires. Consumers increasingly engage information through apps rather than through a common worldwide web. Known also as "Splinternet," regional divides, cultural preferences, and conflicting regulatory environments now shape how and what people experience on the internet.[42] How can democracies and their citizens stick together when they are in different, competing worlds?

Those with advanced tech, time, and resources, what I refer to as Advanced Persistent Manipulators, will increasingly seek to move their audiences to apps of their own design and control. These apps will offer an exponential advantage for harnessing a social media nation as users will either unknowingly or willingly provide manipulators with their personal data when using the app. They'll also be fully trapped in a content bubble where they'll be preyed upon by hidden influencers. Moreover, users will essentially become a human bot network for spreading and amplifying the messages of manipulators. This increased decentralization will harden preference bubbles, further polarization, and create competing worldviews that will be difficult to resolve through cooperation and compromise.

Republicans already openly push for their own apps to bring supporters to an online world of their own design. The National Rifle Association, the pro-Trump political action committee America First, and Senator Ted Cruz's Cruz Crew seek a conservative social media world similar to their Fox News television universe.[43] Curiously, Cambridge Analytica cofounder and former Trump adviser Steve Bannon not only echoes these sentiments but called for nationalizing Facebook.[44] For me, this seems like an odd position for a voracious capitalist who seeks the end of the administrative state. Taking down the tech giants opens enormous space for the strongest manipulators to take hold of unwitting minds; Bannon would be one of those best positioned to gain from their demise.

Disinformation tracking efforts continue to grow and expand. Amazing social media sleuthing by the likes of the online collective BellingCat have restored truth among Russia's disinformation storms. Bot tracking and troll outing has become a pastime for social media enthusiasts around the world, and exhaustive studies of the Internet Research Agency and Iranian Twitter data arise nearly every week. The massive swing of interest toward what my colleagues and I began researching five years ago has been rewarding; it turns out that people do care about Russian trolls after all.

As 2019 kicks off, I'm tracking new and different enemies, looking at how to mess with them, and strangely they are more domestic rather than foreign. American leadership abroad has declined, but I believe American activism has resurged. This book may seem pessimistic in many respects, and I've tried to offer things citizens of democracies can do to turn the tide. Moving forward, though, I feel a touch of optimism about the future, and wonder if our recent social media blow from afar, and our battle within, may be just the shot in the arm America and the Western world needed to determine what they believe in. All is not lost, and while we as citizens may have strayed off course for a bit, I'm confident we can rise again as a better and newer form of good empowered rather than hampered by our digital connections.

December 26, 2018

Acknowledgments

In this book, and throughout my life, a small part of which is captured here, I am deeply indebted to the family, team, and colleagues I've worked with. I owe thanks to two broad ranges of people: those who keep me in line and on track, and those who've helped me serve my country and track and engage my enemies. I'll begin with the first.

No one has taught me more about myself than my daughter, Pepper. Her smiles and hugs are genuine, and her love is unconditional. Thank you Pepper for getting me to look at the world from a different perspective, one that's been invaluable for understanding the new world of social media, which for all its connections is surprisingly isolating.

I was blessed from birth with an amazingly supportive family—my parents, Ronna and Gary, and my sister, Wendy. They taught me right from wrong, stood by me through all the twists and turns in my life, and gave me unending support during life's roughest patches. In this book, I've hopefully answered part of the question "What do you do exactly?" Thank you for sticking with me when you didn't really know the answer to that question.

I'm not sure I would have ever written publicly about Russia's influencing of Americans in the lead up to the presidential election of 2016 if it weren't for a conversation I had with Emmy on a New Jersey sidewalk in July of 2016. Emmy helped me gain the courage to write that first article in the face of smear campaigns, and has since provided unending support and constructive input for this book, something I desperately needed as a first-time author. I can't thank you enough for being there for me.

Thanks to my 1995 West Point classmates and the members of the Long Gray Line. They've been an amazing network I can count on any time and any place I've landed in the world. They are the greatest Americans I've ever met and they all continue to serve their country and their communities across the United States and overseas. Thanks to John for his support on the wisdom of outliers and Jim for being a great teammate in and out of the army.

Thanks to Krista for her support during my years of projects, travel, turn-ins, and short timelines. And much appreciation to Julie for keeping me on track and on target as I navigated the ups and downs of messing with my enemies.

I owe much appreciation to my teachers and professors at Fort Zumwalt North High School in O'Fallon, Missouri, the U.S. Military Academy at West Point, and the Middlebury Institute of International Studies at Monterey, California—particularly Phil Morgan, Fernando DePaolis, Glynn Wood, and Ed Laurance, who equipped me with the skills needed for tackling the research in this book.

Thanks to the members of the Richard A. Pack Society and their families, who've served as a reliable cadre of supporters that continue to defend the country as either public servants or private citizens. You've all been there for me when I needed you the most, and I hope I can return the favor to each of you someday.

Great thanks goes to the two Irish Will's I've come to know: one in the Army and one after. Will McDonough kept me out of trouble and has been a great friend and confidant from my first days at West Point up through now. He, Pat, and Jack (Team 621) have all shown up for me in a big way in recent years. Will McCants and I have worked together since 2005. He's been an amazing supporter and sounding board for my best and worst research ideas. In 2007, I joked about writing a book with Will while sitting alongside the Nile

River. Will stayed on me to see it through, urged me to start blogging at SelectedWisdom.com, provided excellent feedback over the years, and kept me going when times were particularly bleak. Thank you for making sure that I didn't quit.

You may have noticed during the past few hundred pages I've had a meandering career, and I often took the path less traveled. Two leaders recognized how to leverage my crazy, understood how I thought, and encouraged and empowered me to achieve when most others would have left me in the wilderness:

I owe tremendous gratitude to General Wayne A. Downing, the first commander of U.S. Special Operations Command. General Downing was the first general I met who displayed humility, listened more than he talked, and sought to empower those around him. General Downing got me motivated, knew how to nudge me with timely anecdotes, and always encouraged me to push the edges, take on some risk, and march to the sound of the guns, not away from them.

Tom Harrington did what I imagine few FBI leaders would ever do. He not only brought me back to the FBI after I quit, but also empowered me in the Counterterrorism Division, providing an amazing second experience in the Bureau—the one I thought I'd have when I first joined. Tom has been a behind-the-scenes hero in America's counterterrorism fight and led the rebuilding of the FBI into an intelligence-led organization. He mentored me through tough times personally and professionally, and I can't thank him enough for the opportunities he's given me.

Two other characters who've been essential to the research in this book are J. M. Berger and Andrew Weisburd. J.M. is one of the best analysts of social media, terrorism, disinformation, dystopian fiction, and the television show *Lost* in the entire world. J.M.'s relentless quest for novel insights from deep data dives made much of this book possible. Andrew is the best digital shoe-leather investigator I've ever met. I'm constantly amazed by the length and breadth of his research. Andrew's tireless commitment to pursuing bad guys doing bad things has been truly inspiring to me over the last decade. I thank both of them for sticking with the Russia research even when it didn't make any practical sense to pursue it anymore. It's been an honor to work with you both.

I owe my colleagues during my time at the Combating Terrorism Center at West Point a great deal of gratitude. Kip, Joe, Bill, Squeeze, Afshon, Brian, Lianne, Jake, Sammy, Vahid, Becky, Hellfire, Andrew, Jeff, Rick, and all of the outside researchers and staff of the Department of Social Sciences who contributed to great insights in the field of counterterrorism and powered the Harmony projects.

Thanks to the men of 2nd Battalion, 327th Infantry, 101st Airborne Division and 2nd Battalion, 3rd Infantry Regiment, 2nd Infantry Division for letting me learn while leading, particularly the noncommissioned officers—Jason, Scooter, Mike, Shane, and Jeff. Same to the members of the FBI's Counterterrorism Division and National Security branch during my second go-around in the "Bu," and the FBI's Portland Joint Terrorism Task Force, Squad 9, who taught me so much during my first stint—hopefully you all know that while I didn't stay long, I never went too far away.

Thanks to my agent, Flip Brophy, and my publisher, HarperCollins. Both took a gamble on me when many others had passed. Thank you for supporting my concept and for giving me such an excellent editorial team in Jonathan Jao and Sofia Groopman, who gave me essential feedback and got this manuscript up to speed.

Lastly and maybe most strangely, thanks to *The Howard Stern Show*. After more dark days than I care to count, either watching terrorist posts or Russian trolling, you gave me a chuckle when I needed it the most.

Notes

CHAPTER 1: OMAR AND CARFIZZI

1. The original videos of Omar Hammami's raps have been removed and were previously at the following link: https://www.youtube.com /watch?v=MDDYqV9V5kA. A transcript of the rap song that includes this lyric was obtained from J. M. Berger, which summarized the Al Shabaab, Al-Kataib video production entitled "Ambush At Bardale."

2. Clint Watts and Andrew Lebovich, "Hammami's Plight Amidst Al-Shabaab and al-Qaeda's Game of Thrones," Center for Cyber and Homeland Security, *Commentary* 25, (March 19, 2012). https://cchs.gwu.edu /sites/cchs.gwu.edu/files/downloads/Commentary_25_HSPI.pdf.

3. J. M. Berger, "Omar and Me," *Foreign Policy* (September 17, 2013). http://foreignpolicy.com/2013/09/17/omar-and-me.

CHAPTER 2: THE RISE AND FALL OF THE VIRTUAL CALIPHATE

1. Peter Bergen, "The Account of How We Nearly Caught Osama Bin Laden in 2001," *The New Republic*, December 30, 2009. https://new republic.com/article/72086/the-battle-tora-bora.

2. Harmony database, document no. AFGP-2002-6000053, Combating Terrorism Center, West Point, NY, p. 36. https://www.ctc.usma.edu/.

3. Harmony database, document no. AFGP-2002-600113, Combating Terrorism Center, West Point, NY, p. 6. https://www.ctc.usma.edu/.

4. "Exclusive Osama Bin Laden–First Ever TV Interview," CNN Interview conducted by Peter Arnett, Peter Bergen, and Peter Jouvenal. Published on YouTube on January 10, 2012. https://www.youtube.com/watch?v=dqQwnqjA-6w.

5. Anne Stenersen, "The History of the Jihadi Forums," *Jihadica*, March 4, 2009. http://www.jihadica.com/the-history-of-the-jihadi-forums.

6. Hanna Rogan, "Al-Qaeda's Online Media Strategies: From Abu Reuter to Irhabi 007," Norwegian Defence Research Establishment report, December 1, 2007, 48; Brynjar Lia, "Jihadi Web Media Production: Characteristics, Trends, and Future Implications," paper presented at Check the Web conference, Berlin, February 26–27, 2007, 10.

7. Thomas Hegghammer, *Jihad in Saudi Arabia: Violence and Pan-Islamism Since 1979* (Cambridge, UK: Cambridge University Press, 2010), 171–72.

8. Anti-Defamation League, Jihad Online, Islamic Terrorists and the Internet (2002), 14.

9. Brynjar Lia, "Jihadi Web Media Production: Characteristics, Trends, and Future Implications," paper presented in Berlin on February 26, 2007, entitled "Monitoring, Research and Analysis of Jihadist Activities on the Internet—Ways to Deal With the Issue," conference hosted by the German Ministry of Interior, 7–10. https://docs.google.com/viewer?a=v&pid=sites&srcid=ZGVmYXVsdGRvbWFpbnxqaWhhZGlzbXN0dWRpZXNuZXR8Z3g6NTNiOTQ5ZmZmMThkZWQzMA.

10. Rogan, "Al-Qaeda's Online Media Strategies," 65; Lia, "Jihadi Web Media Production," 12.

11. Manuel Ricardo Torres-Soriano, "The Dynamics of the Creation, Evolution, and Disappearance of Terrorist Internet Forums," *International Journal of Conflict and Violence* 7, no. 1 (2013): 175.

12. Maura Conway and Lisa McInerney, "Jihadi Video and Auto-Radicalisation: Evidence from an Exploratory YouTube Study," in *Intelligence and Security Informatics*, ed. D. Ortiz-Arroyo et al. (Berlin and Heidelberg: Springer-Verlag, 2008).

13. Jane Perlez, Salman Masood, and Waqar Gillani, "5 U.S. Men Arrested in Pakistan Said to Plan Jihad," *The New York Times* (December 10, 2009). http://www.nytimes.com/2009/12/11/world/asia/11inquire.html.

14. Andrew Liepman, "Violent Islamist Extremism: Al-Shabaab Recruit-

ment in America," hearing before the Senate Homeland Security and
Governmental Affairs Committee, March 11, 2009.

15. Seth G. Jones, "Awlaki's Death Hits al-Qaeda's Social Media Strategy,"
RAND (September 30, 2011). https://www.rand.org/blog/2011/09
/awlakis-death-hits-al-qaedas-social-media-strategy.html; and Paula New-
ton, "Purported al-Awlaki Message Calls for Jihad Against U.S.," CNN
(March 17, 2010). http://www.cnn.com/2010/WORLD/europe/03/17
/al.awlaki.message/index.html.

16. Dina Temple-Raston, "Grand Jury Focuses on N.C. Man Tied to Jihad
Magazine," *NPR Morning Edition*, (August 18, 2010). https://www.npr
.org/templates/story/story.php?storyId=129263809.

17. Thomas Joscelyn, "Analysis: al Qaeda Groups Reorganize in West Africa,"
Long War Journal (March 13, 2017). https://www.longwarjournal.org
/archives/2017/03/analysis-al-qaeda-groups-reorganize-in-west-africa.php.

18. The single best resource on the breakup of ISIS and al-Qaeda's Nusra is:
William McCants, *The ISIS Apocalypse* (New York: St. Martins Press,
2015). Particularly Chapter 4, "Resurrection and Tribulation," p. 73–98.

19. Clint Watts, "al-Qaeda Plots, NSA Intercepts & the Era of Terrorism
Competition," Foreign Policy Research Institute (August 5, 2013).
https://www.fpri.org/2013/08/al-qaeda-plots-nsa-intercepts-the-era-of-
terrorism-competition.

20. Clint Watts, "Jihadi Competition After al Qaeda Hegemony–The 'Old
Guard', Team ISIS & the Battle for Jihadi Hearts & Minds," Foreign Policy Research Institute (February 20, 2014). https://www.fpri.org/2014/02
/jihadi-competition-after-al-qaeda-hegemony-the-old-guard-team-isis-the
-battle-for-jihadi-hearts-minds.

21. Clint Watts, "ISIS's Rise After al Qaeda's House of Cards," Foreign Policy
Research Institute (March 22, 2014). https://www.fpri.org/2014/03/isiss
-rise-after-al-qaedas-house-of-cards-part-4-of-smarter-counterterrorism.

22. The best reference for ISIS and the Islamic State's social media operations
is: Jessica Stern and J. M. Berger, *ISIS: The State of Terror* (New York:
Ecco, 2015). Particularly chapter 6, "Jihad Goes Social," p. 127–46, and
chapter 7, "The Electronic Brigades," p. 147–76.

23. "2015 Charlie Hebdo Attacks Fast Facts," CNN (December 25, 2017).
http://www.cnn.com/2015/01/21/europe/2015-paris-terror-attacks-fast
-facts/index.html.

CHAPTER 3: "THAT IS NOT AN OPTION UNLESS IT'S IN A BODY BAG"

1. "Al-Shabaab Militants in Somalia Post Alleged Photo of French Commando Killed in Botched Raid," CBSnews.com (January 14, 2013). https://www.cbsnews.com/news/al-shabab-militants-in-somalia-post-alleged-photo-of-french-commando-killed-in-botched-raid.

2. Harmony database, document no. AFGP-2002-800640, Combating Terrorism Center, West Point, NY, p. 7. https://www.ctc.usma.edu/.

3. Ken Menkhaus, "al-Shabaab and Social Media: A Double-Edged Sword," *The Brown Journal of World Affairs* (Spring/Summer 2014). https://www.brown.edu/initiatives/journal-world-affairs/sites/brown.edu.initiatives.journal-world-affairs/files/private/articles/Menkhaus.pdf.

4. "Kenya's Military Tweets of Latest Threat in Somalia War: Donkeys," NBC News (November 3, 2011). http://www.nbcnews.com/id/45154226/ns/world_news-africa/t/kenyas-military-tweets-latest-threat-somalia-war-donkeys/#.WcleB0yZNTY.

5. David Smith, "Al-Shabaab in War of Words with Kenyan Army on Twitter," *The Guardian* (December 13, 2011). https://www.theguardian.com/world/2011/dec/13/al-shabaab-war-words-twitter.

6. "'American al-Shabab' Disavows Militant Group, al-Qaida," Voice of America (September 5, 2013). https://www.voanews.com/a/american-alshabab-disavows-militant-group-alqaida/1743983.html.

CHAPTER 4: RISE OF THE TROLLS

1. Mike Krumboltz, "WhiteHouse.gov Petition Seeks to Give Alaska Back to Russia," Yahoo (March 27, 2014). https://www.yahoo.com/news/blogs/sideshow/whitehouse-gov-petition-seeks-to-give-alaska-back-to-russia-165159170.html.

2. Edmund Lee, "AP Twitter Account Hacked in Market-Moving Attack," *Bloomberg Business* (April 23, 2013). https://www.bloomberg.com/news/articles/2013-04-23/dow-jones-drops-recovers-after-false-report-on-ap-twitter-page.

3. Max Fisher, "Syrian hackers Claim AP Hack That Tipped Stock Market by $136 billion. Is It Terrorism?," *The Washington Post* (April 23, 2013).

https://www.washingtonpost.com/news/worldviews/wp/2013/04/23
/syrian-hackers-claim-ap-hack-that-tipped-stock-market-by-136
-billion-is-it-terrorism/?utm_term=.0cb10e61e8fc; James Temperton, "FBI
Adds Syrian Electronic Army Hackers to Most Wanted List," *Wired*
(March 23, 2016). http://www.wired.co.uk/article/syrian-electronic-army
-fbi-most-wanted.

4. For a short summary of the "Turing Test", Wikipedia does a good break-
    down. https://en.wikipedia.org/wiki/Turing_test.

5. Phil Howard, "Computational Propaganda: The Impact of Algorithms
    and Automation on Public Life," Presentation available at: https://prezi
    .com/b_vewutjwzut/computational-propaganda/?webgl=0.

6. Caitlin Dewey, "One in Four Debate Tweets Comes from a Bot. Here's
    How to Spot Them," *The Washington Post* (October 19, 2016). https://
    www.washingtonpost.com/news/the-intersect/wp/2016/10/19/one-in
    -four-debate-tweets-comes-from-a-bot-heres-how-to-spot-them.

7. Manny Fernandez, "Conspiracy Theories Over Jade Helm Training Ex-
    ercise Get Some Traction in Texas," *The New York Times* (May 6, 2015).
    https://www.nytimes.com/2015/05/07/us/conspiracy-theories-over
    -jade-helm-get-some-traction-in-texas.html.

8. For examples of Jade Helm posts, here are some representative articles:
    "Jade Helm 15: What You Need to Know About 7-State Pentagon Su-
    per-Drill," RT (July 14, 2015). https://www.rt.com/usa/273436-jade
    -helm-drill-explained; Robert Bridge, "Jade Helm 15: One nation under
    siege?," RT (July 10, 2015). https://www.rt.com/op-edge/272920-us-army
    -jade-helm; "Are Jade Help 15 Drills Disguise to Enact Martial Law
    in US?," Sputnik (May 27, 2015). https://sputniknews.com/us/2015
    05271022628702; and "Jade Help 15: Texans Terrified of Obama-led
    US Army Invasion," Sputnik (July 7, 2015). https://sputniknews.com
    /us/201507071024303072.

9. Wade Goodwyn, "Texas Governor Deploys State Guard to Stave Off
    Obama Takeover," NPR (May 2, 2015). https://www.npr.org/sections
    /itsallpolitics/2015/05/02/403865824/texas-governor-deploys-state
    -guard-to-stave-off-obama-takeover.

10. Adrian Chen, "The Agency," *The New York Times* (June 2, 2015). https://
    www.nytimes.com/2015/06/07/magazine/the-agency.html.

CHAPTER 5: HARMONY, DISHARMONY, AND THE POWER OF SECRETS

1. "Translation of Aweis Letter.doc," Wikileaks (December 28, 2006). https://wikileaks.org/wiki/Translation_of_Aweis_Letter.doc.

2. "Top Somali Islamist Flown to Mogadishu 'After Split,' " BBC News (June 29, 2013). http://www.bbc.com/news/world-africa-23115819.

3. Raffi Khatchadourian, "No Secrets," The New Yorker (June 7, 2010). https://www.newyorker.com/magazine/2010/06/07/no-secrets.

4. Ibid.

5. Shaun Walker, "Was Russian Secret Service Behind Leak of Climate-Change Emails?," Independent (December 7, 2009). http://www.independent.co.uk/news/world/europe/was-russian-secret-service-behind-leak-of-climate-change-emails-1835502.html.

6. Fred Weir, "WikiLeaks Ready to Drop a Bombshell on Russia. But Will Russians Get to Read About It?," The Christian Science Monitor (October 26, 2010). https://www.csmonitor.com/World/Europe/2010/1026/WikiLeaks-ready-to-drop-a-bombshell-on-Russia.-But-will-Russians-get-to-read-about-it.

7. Simon Shuster, "Wikileaks. Is Russia the Next Target?," Time (November 1, 2010). http://content.time.com/time/world/article/0,8599,2028283,00.html.

8. James Ball, "Why I Felt I Had to Turn my Back on Wikileaks," The Guardian (September 2, 2011). https://www.theguardian.com/media/2011/sep/02/why-i-had-to-leave-wikileaks.

9. Kapil Komireddi, "Julian Assange and Europe's Last Dictator, Alexander Lukashenko," New Statesman (March 1, 2012). https://www.newstatesman.com/blogs/the-staggers/2012/03/belarus-assange-lukashenko.

10. "U.S.–Led Attack on Afghanistan Begins," History.com (2010). http://www.history.com/this-day-in-history/u-s-led-attack-on-afghanistan-begins.

11. Gregory D. Johnsen, The Last Refuge: Yemen, al-Qaeda, and America's War in Arabia., W.W. Norton (November 19, 2012). https://www.amazon.com/Last-Refuge-Al-Qaeda-Americas-Arabia/dp/0393082423.

12. "Document and Media Exploitation (DOMEX)," U.S. Army (2009). https://www.army.mil/aps/09/information_papers/document_media_exploitation.html.

13. An overview of the Combating Terrorism Center's Harmony Program

can be found at their website. https://ctc.usma.edu/programs-resources /harmony-program.

14. Joseph Felter, Jacob Shapiro, et al., *Harmony and Disharmony: Exploiting al-Qa'ida's Organizational Vulnerabilities*, Combating Terrorism Center, (February 14, 2006). https://ctc.usma.edu/programs-resources/harmony -program.

15. Ibid, 42.

16. Ibid, 43.

17. Ibid, 43.

18. Clint Watts, Jacob Shapiro, Vahid Brown, et al., *al-Qa'ida's (Mis)Adventures in the Horn of Africa*, Combating Terrorism Center (2007). https:// ctc.usma.edu/al-qaidas-misadventures-in-the-horn-of-africa/.

19. Vahid Brown, *Cracks in the Foundation: Leadership Schisms in al-Qa'ida From 1989–2006*, Combating Terrorism Center (2007). https://ctc .usma.edu/cracks-in-the-foundation-leadership-schisms-in-al-qaida -from-1989-2006/.

20. Ibid.

21. A copy of Sheikh Hassan Aweys's Kenya visa application can be found here: https://ctc.usma.edu/app/uploads/2013/10/Hassan-Aweys-Kenya -Visa-Application-Translation.pdf.

22. James R. Hollyer, B. Peter Rosendorff and James Raymond Vreeland, "Democracy and Transparency," *The Journal of Politics*, vol. 73, No. 4 (Oct. 21, 2011), 1191–1205. https://wp.nyu.edu/faculty-rosendorff/wp -content/uploads/sites/1510/2015/03/HRVJOP.pdf.

23. Eric Schmitt, "C.I.A. Warning on Snowden in '09 Said to Slip Through the Cracks," *The New York Times* (October 10, 2013). http://www.ny times.com/2013/10/11/us/cia-warning-on-snowden-in-09-said-to-slip -through-the-cracks.html.

24. "Executive Summary of Review of the Unauthorized Disclosures of Former National Security Agency Contractor Edward Snowden," U.S. House of Representatives (September 15, 2016). https://intelligence.house.gov /uploadedfiles/hpsci_snowden_review-unclasssummary-final.pdf.

25. An edited extract of a copy of Julian Assange's "Conspiracy as Governance," was posted by the Frontline Club on June 28, 2011 and is available at: https://www.frontlineclub.com/julian_assange_the_state_and_terror ist_conspiracies/.

26. Khatchadourian, "NoSecret."

27. Ibid.

28. Mariano Castillo, "Bodies Hanging from Bridge in Mexico Are Warning to Social Media Users," CNN (September 15, 2011). http://www.cnn.com/2011/WORLD/americas/09/14/mexico.violence/index.html.

29. Ibid.

30. For a summary of the Panama Papers see Will Fitzgibbon and Emilia Diaz-Struck, "The Panama Papers," International Consortium of Investigative Journalists (December 1, 2016). https://panamapapers.icij.org/blog/20161201-impact-graphic.html.

31. Juliette Garside, Holly Watt, and David Pegg, "The Panama Papers: How the World's Rich and Famous Hide Their Money Offshore," *The Guardian* (April 3, 2016). https://www.theguardian.com/news/2016/apr/03/the-panama-papers-how-the-worlds-rich-and-famous-hide-their-money-offshore.

32. Luke Harding, "Revealed: The $2bn Offshore Trail that Leads to Vladimir Putin," *The Guardian* (April 3, 2016). https://www.theguardian.com/news/2016/apr/03/panama-papers-money-hidden-offshore.

33. Michael Forsythe, "Paradise Papers Shine Light on Where the Elite Keep Their Money," *The New York Times* (November 5, 2017). https://www.nytimes.com/2017/11/05/world/paradise-papers.html.

34. "Daphne Caruana Galizia: The Blogging Fury," *Politico* (2017). https://www.politico.eu/list/politico-28-class-of-2017-ranking/daphne-caruana-galizia.

35. Juliette Garside, "Malta Car Bomb Kills Panama Papers Journalist," *The Guardian* (October 16, 2017). https://www.theguardian.com/world/2017/oct/16/malta-car-bomb-kills-panama-papers-journalist.

36. "Daphne Caruana Galizia: The Blogging Fury."

37. Andrei Soldatov and Irina Borogan, *The Red Web: The Struggle Between Russia's Digital Dictators and the New Online Revolutionaries* (New York: PublicAffairs, 2015), 215.

CHAPTER 6: PUTIN'S PLAN

1. Thomas Boghardt, "Active Measures: The Russian Art of Disinformation," AIRSHO (October 2006), 20–26. https://spy-museum.s3.amazonaws.com/files/back_active-measures.pdf.

2. "History of HIV and AIDS," Avert.org (March 9, 2018). https://www
    .avert.org/professionals/history-hiv-aids/overview.

3. Thomas Boghardt, "Operation INFEKTION: Soviet Bloc Intelli-
    gence and Its AIDS Disinformation Campaign," *Studies in Intelligence*,
    vol. 53, No. 4. (December 2009) 1–24. https://www.cia.gov/library
    /center-for-the-study-of-intelligence/csi-publications/csi-studies
    /studies/vol53no4/pdf/U-%20Boghardt-AIDS-Made%20in%20
    the%20USA-17Dec.pdf.

4. Robert Coalson, "Top Russian General Lays Bare Putin's Plan for
    Ukraine," *The Huffington Post* (November 2, 2014). https://www.huffing
    tonpost.com/robert-coalson/valery-gerasimov-putin-ukraine_b_574
    8480.html.

5. "Russia to create 'cyber-troops'—Ministry of Defense," RT (August 20,
    2013). https://www.rt.com/news/russia-cyber-troops-defence-753.

6. Eric Lipton, David E. Sanger, and Scott Shane, "The Perfect Weapon:
    How Russian Cyberpower Invaded the U.S.," *The New York Times* (De-
    cember 13, 2016). https://www.nytimes.com/2016/12/13/us/politics
    /russia-hack-election-dnc.html.

7. Joe Uchill, "Typo Led to Podesta Email Hack: Report," *The Hill* (De-
    cember 13, 2016). http://thehill.com/policy/cybersecurity/310234-typo
    -may-have-caused-podesta-email-hack.

8. Lipton, Sanger and Shane, "The Perfect Weapon."

9. Patrick Tucker, "EXCLUSIVE: Russia-Backed DNC Hackers Strike
    Washington Think Tanks," *Defense One* (August 29, 2016). http://www
    .defenseone.com/threats/2016/08/exclusive-russia-backed-dnc-hackers
    -strike-washington-think-tanks/131104.

10. Massimo Calabresi, "Inside Russia's Social Media War on America,"
    Time (May 18, 2017). http://time.com/4783932/inside-russia-social-me-
    dia-war-america.

11. "Is Donald Trump a Man to Mend US Relations with Russia?," Sput-
    nik (August 24, 2015). https://sputniknews.com/politics/201508241
    026144021.

12. David Ignatius, "Russia's Radical New Strategy for Information Warfare,"
    The Washington Post (January 18, 2017). https://www.washingtonpost
    .com/blogs/post-partisan/wp/2017/01/18/russias-radical-new-strategy
    -for-information-warfare/?utm_term=.1873e03e1c47.

13. Ashley Parker and David Sanger, "Donald Trump Calls on Russia to Find Hillary Clinton's Missing Emails," *The New York Times* (July 27, 2016). https://www.nytimes.com/2016/07/28/us/politics/donald-trump-russia -clinton-emails.html.

14. Clint Watts and Andrew Weisburd, "How Russia Dominates Your Twitter Feed to Promote Lies (And, Trump, Too)," *The Daily Beast* (August 6, 2016). https://www.thedailybeast.com/how-russia-dominates-your-twitter -feed-to-promote-lies-and-trump-too.

15. Ryan Goodman, "How Roger Stone Interacted With Russia's Guccifer and Wikileaks," *Newsweek* (September 28, 2017). http://www.news week.com/how-stone-interacted-russias-guccifer-and-wikileaks-673268.

16. Ibid.

17. Matthew Cole, Richard Esposito, Sam Biddle, and Ryan Grim, "Top-Secret NSA Report Details Russian Hacking Effort Days Before 2016 Election," *The Intercept* (June 5, 2017). https://theintercept.com/2017/06/05/top-secret-nsa-report-details-russian-hacking-effort-days-before-2016-election.

18. Bence Kollanyi, Philip N. Howard, and Samuel C. Woolley. "Bots and Automation over Twitter during the Third U.S. Presidential Debate." Data Memo 2016.3, Oxford, UK: Project on Computational Propaganda (October 31, 2016). http://comprop.oii.ox.ac.uk/research/working-papers/ bots-and-automation-over-twitter-during-the-third-u-s-presidential-debate.

CHAPTER 7: POSTMORTEM

1. Craig Silverman, "This Analysis Shows How Viral Fake Election News Stories Outperformed Real News On Facebook," *Buzzfeed News* (November 16, 2016). https://www.buzzfeed.com/craigsilverman/viral-fake -election-news-outperformed-real-news-on-facebook?utm_term=.fw 8MNAMD0#.wf9qNQqdz.

2. Jacob Poushter, "Not Everyone in Advanced Economies Is Using Social Media," Pew Research Center (April 20, 2017). http://www.pewresearch .org/fact-tank/2017/04/20/not-everyone-in-advanced-economies-is -using-social-media.

3. Yochai Benkler, Robert Faris, Hal Roberts, and Ethan Zuckerman, "Study: Breitbart-Led Right-Wing Media Ecosystem Altered Broader

Media Agenda," *Columbia Journalism Review* (March 3, 2017). https://www.cjr.org/analysis/breitbart-media-trump-harvard-study.php.

4. Ned Parker, Jonathan Landay, and John Walcott, "Putin-Linked Think Tank Drew Up Plan to Sway 2016 US Election—Documents," Reuters (April 19, 2017). http://www.reuters.com/article/us-usa-russia-election -exclusive-idUSKBN17L2N3.

5. "Digital Nation Data Explorer," National Telecommunications and Information Administration (October 27, 2016). https://www.ntia.doc .gov/data/digital-nation-data-explorer#sel=internetUser&disp=map.

6. "Presidential Results," CNN (Accessed March 19, 2018). http://www .cnn.com/election/results/president.

7. Chengcheng Shao, Giovanni Luca Ciampaglia, Onur Varol, Alessandro Flammini, and Filippo Menczer, "The Spread of Fake News by Social Bots," Indiana University, Bloomington (July 24, 2017). https://www .researchgate.net/publication/318671211_The_spread_of_fake_news _by_social_bots.

8. Mike Isaac and Daisuke Wakabayashi, "Russian Influence Reached 126 Million Through Facebook Alone," *The New York Times* (October 30, 2017). https://www.nytimes.com/2017/10/30/technology/face book-google-russia.html.

9. Donie O'Sullivan, "Russian Trolls Created Facebook Events Seen by More than 300,000 Users," CNNMoney (January 26, 2018). http://money.cnn .com/2018/01/26/media/russia-trolls-facebook-events/index.html.

10. Rob Price, "Google Has Detailed How Russia Tried to Use Its Platforms to Influence US Politics," *Business Insider* (October 31, 2017). http://www .businessinsider.com/google-report-russian-election-meddling-us-con gress-testimony-2017-10.

11. Spencer Ackerman, Gideon Resnick, and Ben Collins, "Exclusive: Russia Recruited YouTubers to Bash 'Racist B*tch' Hillary Clinton Over Rap Beats," *The Daily Beast* (August 10, 2017). https://www.thedailybeast .com/russia-recruited-youtubers-to-bash-racist-btch-hillary-clinton-over -rap-beats.

12. Olivia Solon and Sabrina Siddiqui, "Russia-Backed Facebook Posts 'Reached 126m Americans' during US Election," *The Guardian* (October 30, 2017). https://www.theguardian.com/technology/2017/oct/30 /facebook-russia-fake-accounts-126-million.

13. Ben Collins, Kevin Poulsen, Spencer Ackerman, and Betsy Woodruff, "Trump Campaign Staffers Pushed Russian Propaganda Days Before the Election," *The Daily Beast* (October 18, 2017). https://www.thedaily beast.com/trump-campaign-staffers-pushed-russian-propaganda-days -before-the-election.

14. Kevin Collier, "Prominent "GOP" Twitter Account, Allegedly a Russian Troll, Was Widely Quoted in US Media," *Buzzfeed* (October 18, 2017). https://www.buzzfeed.com/kevincollier/americans-helped-spread-an -alleged-russian-gop-accounts?utm_term=.xd6GxOGVj#.atE3z53VQ.

15. Robert Windrem, "Guess Who Came to Dinner With Flynn and Putin," NBC News (April 18, 2017). http://www.nbcnews.com/news/world /guess-who-came-dinner-flynn-putin-n742696.

16. Josh Rogin, "Trump Campaign Guts GOP's Anti-Russia Stance on Ukraine," *The Washington Post* (July 18, 2016). https://www.washington post.com/opinions/global-opinions/trump-campaign-guts-gops-anti -russia-stance-on-ukraine/2016/07/18/98adb3b0-4cf3-11e6-a7d8-13d06 b37f256_story.html.

17. Jack Gillum, Chad Day, and Jeff Horwitz, "Manafort Firm Received Ukraine Ledger Payout," Associated Press (April 12, 2017). https://www .apnews.com/20cfc75c82eb4a67b94e624e97207e23.

18. Elliot Hannon, "U.S. Reportedly Intercepted Suspected Russian Agents' Chatter That Manafort Asked for Their Help With Clinton," *Slate* (August 3, 2017). http://www.slate.com/blogs/the_slatest/2017/08/03/u_s_ intelligence_reportedly_intercepted_russian_operatives_chatter_about.html.

19. Carol D. Leonnig, Tom Hamburger, and Rosalind S. Helderman, "Trump's Business Sought Deal on a Trump Tower in Moscow While He Ran for President," *The Washington Post* (August 27, 2017). https://www .washingtonpost.com/politics/trumps-business-sought-deal-on-a-trump -tower-in-moscow-while-he-ran-for-president/2017/08/27/d6e95114 -8b65-11e7-91d5-ab4e4bb76a3a_story.html.

20. Adam Goldman, "Russian Spies Tried to Recruit Carter Page Before He Advised Trump," *The New York Times* (April 4, 2017). https://www.ny times.com/2017/04/04/us/politics/carter-page-trump-russia.html.

21. "Trump Tower Russia meeting: At least eight people in the room," CNN (July 15, 2017). http://www.cnn.com/2017/07/14/politics/donald-trump -jr-meeting/index.html.

22. "Read the emails on Donald Trump Jr.'s Russia Meeting," *The New York Times* (July 11, 2017). https://www.nytimes.com/interactive/2017/07/11/us/politics/donald-trump-jr-email-text.html.

23. Euan McKirdy and Mary Ilyushina, "Putin: 'Patriotic' Russian Hackers May Have Targeted US Election," CNN (June 2, 2017). http://www.cnn.com/2017/06/01/politics/russia-putin-hackers-election/index.html.

24. Max Kutner, "Russian Embassy in London Joins in Seth Rich Murder Speculation," *Newsweek* (May 19, 2017). http://www.newsweek.com/russian-embassy-twitter-seth-rich-murder-dnc-612446.

25. Ekaterina Blinova, "Unresolved Murder: Why Seth Rich's Case is Key to #TrumpRussia Investigation," Sputnik (June 2, 2017). https://sputniknews.com/politics/201706021054260081-seth-rich-trump-russia; and Russian embassy, UK (@RussianEmbassy), "#WikiLeaks informer Seth Rich murdered in US but (UK Flag) MSM was so busy accusing Russian hackers to take notice." Twitter, May 19, 2017. https://twitter.com/russianembassy/status/865571158862811136?lang=en.

26. Russian Embassy, UK (@RussianEmbassy), "#Wikileaks informer Seth Rich murdered in US but [UK Flag] MSM was so busy accusing Russian hackers to take notice."

27. Emily Steel, "Lawsuit Asserts White House Role in Fox News Article on Seth Rich," *The New York Times* (August 1, 2017). https://www.nytimes.com/2017/08/01/business/media/seth-rich-fox-news-white-house-lawsuit.html.

28. Michael Schwirtz and Joseph Goldstein, "Russian Espionage Piggybacks on a Cybercriminal's Hacking," *The New York Times* (March 12, 2017). https://www.nytimes.com/2017/03/12/world/europe/russia-hacker-evgeniy-bogachev.html; and Garrett M. Graff, "Inside the Hunt For Russia's Most Notorious Hacker," *Wired* (March 21, 2017). https://www.wired.com/2017/03/russian-hacker-spy-botnet.

29. Павел Седаков, "Контракт со взломом: как хакер построила бизнес за счет банков и корпораций" *Forbes* (November 12, 2014). http://www.forbes.ru/tekhnologii/internet-i-svyaz/275355-kontrakt-na-ugrozu-kak-khaker-stroit-biznes-na-zashchite-bankov. For more on Es-age Lab and Neuron Hackerspace see: http://esagelab.com/neuronspace; and Andy Greenberg, "How an Entire Nation Became Russia's Test Lab For Cyberwar," *Wired* (June 20, 2017). https://www.wired.com/story/russian-hackers-attack-ukraine.

30. Alan Rappeport and Neil MacFarquhar, "Trump Imposes New Sanctions on Russia Over Ukraine Incursion," *The New York Times* (June 20, 2017). https://www.nytimes.com/2017/06/20/world/europe/united-states-sanctions-russia-ukraine.html.

31. Jack Losh, "Putin's Angels: The Bikers Battling for Russia in Ukraine," *The Guardian* (January 29, 2016). https://www.theguardian.com/world/2016/jan/29/russian-biker-gang-in-ukraine-night-wolves-putin.

32. Sarah Lyall, "Spies Like Us: A Conversation with John le Carre and Ben Macintyre," *The New York Times* (August 25, 2017). https://mobile.ny times.com/2017/08/25/books/review/john-le-carre-ben-macintyre-british-spy-thrillers.html.

33. Seva Gunitsky, "Commentary: Fascism Spread in 1930s America. It Could Spread Again Today," *Chicago Tribune* (August 16, 2017). http://www.chicagotribune.com/news/opinion/commentary/ct-trump-fascism-nazis-hitler-0170816-story.html.

34. Amanda Taub, "How Stable Are Democracies? 'Warning Signs Are Flashing Red,' " *The New York Times* (November 29, 2016). https://www.nytimes.com/2016/11/29/world/americas/western-liberal-democracy.html.

35. Amy Erica Smith, "Do Americans Still Believe in Democracy?," *The Washington Post* (April 9, 2016). https://www.washingtonpost.com/news/monkey-cage/wp/2016/04/09/do-americans-still-believe-in-democracy.

36. Ariel Malka and Yphtach Lelkes, "In a New Poll, Half of Republicans Say They Would Support Postponing the 2020 Election if Trump Proposed It," *The Washington Post* (August 10, 2017). https://www.washingtonpost.com/news/monkey-cage/wp/2017/08/10/in-a-new-poll-half-of-republicans-say-they-would-support-postponing-the-2020-election-if-trump-proposed-it.

37. "Torch-wielding protestors gather at Lee Park," *The Daily Progress* (May 13, 2017). http://www.dailyprogress.com/news/local/torch-wielding-protesters-gather-at-lee-park/article_201dc390-384d-11e7-bf16-fb43de0f5d38.html.

38. Dan MacGuill, "Torch-Bearing White Supremacists Rally in Charlottesville, Virginia," Snopes (May 15, 2017). https://www.snopes.com/2017/05/15/torches-rally-charlottesville.

39. Brandon Wall (@Walldo), "'Assad did nothing wrong'"—Baked Alaska at UVA tonight. 'Barrel bombs, hell yeah!'," Twitter (August 11, 2017). https://twitter.com/walldo/status/896212779069186049?lang=en.

40. Liz Sly and Rick Noack, "Syria's Assad Has Become an Icon of the Far Right in America," *The Washington Post* (August 14, 2017). https://www.washingtonpost.com/news/worldviews/wp/2017/08/13/syrias-assad-has-become-an-unexpected-icon-of-the-far-right-in-america/?utm_term=.dd687a3d3e82.

41. Roland Oliphant, "Montenegro Defies Russia to Join Nato as Alliance's 29th Member," *The Telegraph* (June 5, 2017). http://www.telegraph.co.uk/news/2017/06/05/montenegro-joins-nato-alliances-29th-member.

42. Alec Luhn, "Montenegro Finds Itself at Heart of Tensions with Russia As It Joins Nato," *The Guardian* (May 25, 2017). https://www.theguardian.com/world/2017/may/25/montenegro-tensions-russia-joins-nato-member.

43. "Kremlin Rejects Claims Russia Had Role in Montenegro Coup Plot," *The Guardian* (February 20, 2017). https://www.theguardian.com/world/2017/feb/20/russian-state-bodies-attempted-a-coup-in-montenegro-says-prosecutor.

44. Predrag Milic, "Montenegro: 'Nationalists from Russia' Organized Assassination Attempt Against Prime Minister Djukanovic," Associated Press (November 21, 2016). http://www.atlanticcouncil.org/blogs/natosource/montenegro-nationalists-from-russia-organized-assassination-attempt-against-prime-minister-djukanovic.

45. Daniella Diaz, "Watch President Trump Push a Prime Minister Aside," CNN (May 25, 2017). http://www.cnn.com/2017/05/25/politics/trump-pushes-prime-minister-nato-summit/index.html.

46. Melanie Hall, "Russia 'Trying to Destabilise' Germany by Stoking Unrest over Migrants, Warn Spy Chiefs," *The Telegraph* (March 10, 2016). http://www.telegraph.co.uk/news/worldnews/europe/germany/12190353/Russia-trying-to-destabilise-Germany-by-stoking-unrest-over-migrants-warn-spy-chiefs.html.

47. Damien McGuinness, "Russia Steps in Berlin 'Rape' Storm Claiming German Cover-Up," BBC News (January 27, 2016). http://www.bbc.com/news/blogs-eu-35413134.

48. Benjamin Bidder, "Russland Wirft deutschen Behorden Vertuschung vor," *Spiegel* (January 26, 2016). http://www.spiegel.de/politik/ausland/berlin -lawrow-zu-angeblicher-vergewaltigung-von-13-jaehriger-a-1073933 .html; and Adam Withnall, "Russian-German Girl 'Admits Making Up' Claim She Was Raped by Refugees in Berlin," *Independent* (January 31 2016). http://www.independent.co.uk/news/world/europe/russian-german -girl-admits-making-up-claim-she-was-raped-by-refugees-in-berlin -a6845256.html.

49. Alexey Kovalev, "The Secrets of Russia's Propaganda War, Revealed," *The Moscow Times* (March 1, 2017). https://themoscowtimes.com/articles /welcome-to-russian-psychological-warfare-operations-101-57301.

50. Rosie Gray, "The 'Macron Leaks' Rebel in the Briefing Room," *The Atlantic* (May 10, 2017). https://www.theatlantic.com/politics/archive /2017/05/the-macron-leaks-rebel-in-the-briefing-room/526065.

51. Alexander Hamilton, *The Federalist* No. 68 (March 12, 1788), in Alexander Hamilton, James Madison, and John Jay, *The Federalist: A Commentary on the Constitution of the United States*, ed. Robert Scigliano (New York: Modern Library, 2000), 436.

52. "Ex-Russian Ambassador to US Kislyak Surprised by US Efforts to Track Sputnik, RT," Sputnik (August 5, 2017). https://sputniknews.com/us /201708051056210952-kislyak-surprise-us-track-sputnik.

CHAPTER 8: STARING AT THE MEN WHO STARE AT GOATS

1. The contents and select declassified communications recovered from the raid on Osama bin Laden's Abbottabad, Pakistan, compound can be accessed at the Director of National Intelligence website at "Bin Laden's Bookshelf," published on May 20, 2015. https://www.dni.gov/index .php/features/bin-laden-s-bookshelf.

2. Thomas Shanker, "Defense Secretary Urges More Spedning for U.S. Diplomacy," *The New York Times* (November 27, 2007). http://www .nytimes.com/2007/11/27/washington/27gates.html.

3. An overview of the U.S. Information Agency's mission and activities can be found in this 1998 summary: http://dosfan.lib.uic.edu/usia/usia home/oldoview.htm#overview.

4. Jacob Silverman, "The State Department's Twitter Jihad," *Politico* (July 22, 2014). https://www.politico.com/magazine/story/2014/07 /the-state-departments-twitter-jihad-109234.

5. Michael Lumpkin, "As ISIS Evolves, U.S. Counter-Efforts Must Advance, Lumpkin Says", interview by Renee Montagne, *Morning Edition*, NPR, https://www.npr.org/2016/02/01/465106713/as-isis-evolves-u-s -counter-efforts-must-advance-lumpkin-says.

CHAPTER 9: FROM PREFERENCE BUBBLES TO SOCIAL INCEPTION

1. Eli Pariser, *The Filter Bubble: How the New Personalized Web Is Changing What We Read and How We Think*, Penguin Books (April 24, 2012). https://www.amazon.com/Filter-Bubble-Personalized-Changing-Think /dp/0143121235.

2. Tom Nichols, "The Death Of Expertise," *The Federalist* (January 17, 2014) http://thefederalist.com/2014/01/17/the-death-of-expertise.

3. Thomas Jocelyn, "Abu Qatada Provides Jihadists with Ideological Guidance from a Jordanian Prison," *The Long War Journal* (April 30, 2014). https://www.longwarjournal.org/archives/2014/04 /jihadist_ideologue_p.php; and Spencer Ackerman, Shiv Malik, Ali Younes, and Mustafa Khalili, "Al-Qa'ida 'Cut Off and Ripped Apart by Isis,'" *The Guardian* (June 15, 2015). https://www.theguardian. com/world/2015/jun/10/isis-onslaught-has-broken-al-qaida-its- spiritual-leaders-admit.

4. William McCants, "The Believer," Brookings (September 1, 2015). http://csweb.brookings.edu/content/research/essays/2015/thebeliever .html.

5. Seth Borenstein, "Trump science job nominees missing advanced science degrees," Associated Press (December 5, 2017). https://www.apnews .com/fc357285cc2d491abfa5cc0d817603ba.

6. Lauren Etter, "What Happens When the Government Uses Facebook as a Weapon?" *Bloomberg Businessweek* (December 7, 2017). https://www .bloomberg.com/news/features/2017-12-07/how-rodrigo-duterte-turned -facebook-into-a-weapon-with-a-little-help-from-facebook.

CHAPTER 10: SURVIVING IN A SOCIAL MEDIA WORLD

1. Howard Schultz, "Howard Schultz: The Star of Starbucks," interview by Scott Pelley, *60 Minutes*, CBS News, https://www.cbsnews.com/news /howard-schultz-the-star-of-starbucks.

2. Carl Bialik, "Starbucks Stays Mum on Drink Math," *The Wall Street Journal* (April 2, 2008). https://blogs.wsj.com/numbers/starbucks-stays -mum-on-drink-math-309/.

3. Alexis de Tocqueville, *Democracy in America*, translated by Henry Reeve (State College, PA: Penn State University, (2002), 581.

4. "Social Capital Community Benchmark Survey: Preliminary Results," The Saguaro Seminar Civic Engagement in America, JFK School of Government, Harvard University (2001). https://sites.hks.harvard.edu /saguaro/communitysurvey/results3.html.

5. James Wood, "Tocqueville in America: The Grand Journey, Retraced and Reimagined," *The New Yorker* (May 17, 2010). https://www.new yorker.com/magazine/2010/05/17/tocqueville-in-america.

6. "Do Social Media threaten Democracy?" *The Economist*, (November 4, 2017). https://www.economist.com/news/leaders/21730871-facebook-google-and-twitter-were-supposed-save-politics-good-information-drove-out.

7. Allan J. Kimmel and Robert Keefer, "Psychological Correlates of the Transmission of Rumors About AIDS," *Journal of Applied Social Psychology* 21, no. 19 (October 1991), 1608–28. http://onlinelibrary.wiley.com /doi/10.1111/j.1559-1816.1991.tb00490.x/full.

8. Craig Silverman, Lauren Strapagiel, Hamza Shaban, Ellie Hall, and Jeremy Singer-Vine, "Hyperpartisan Facebook Page Publishing False and Misleading Information at an Alarming Rate," *Buzzfeed News* (October 20, 2016). https://www.buzzfeed.com/craigsilverman/partisan-fb-pages -analysis?utm_term=.rmZZVOZpG#.skWnzJnkQ.

9. Robert H. Knapp, "A Psychology of Rumor," *Public Opinion Quarterly* 8, no. 1 (January 1, 1944), 22–7. https://academic.oup.com/poq/article -abstract/8/1/22/1914214?redirectedFrom=fulltext.

10. Craig Silverman, "Lies, Damn Lies and Viral Content: How News Websites Spread (And Debunk) Online Rumors, Unverified Claims and Misinformation," Tow Center for Digital Journalism, Columbia Journalism School (February 15, 2015), 28. https://towcenter.org/wp-content

/uploads/2015/02/LiesDamnLies_Silverman_TowCenter.pdf.

11. Prashant Bordia and Nicholas DiFonzo, *Rumor Psychology: Social and Organizational Approaches* (Washington, D.C.: American Psychological Association, 2007).

12. Adam Alter, *Irresistible: The Rise of Addictive Technology and the Business of Keeping Us Hooked* (New York: Penguin Press, 2017), 28.

13. Peter Walker, "Facebook Makes You Unhappy and Makes Jealous People Particularly Sad, Study Finds," *The Independent* (December 22, 2016). http://www.independent.co.uk/life-style/gadgets-and-tech/facebook -social-media-make-unhappy-jealous-people-particularly-sad-copenhagen -university-study-a7490816.html.

14. Kayleigh Lewis, "Heavy Social Media Users 'Trapped in Endless Cycle of Depression," *The Independent* (March 24 2016). https://www.inde pendent.co.uk/life-style/health-and-families/health-news/social-media -depression-facebook-twitter-health-young-study-a6948401.html.

15. "Have We Created Unsocial Media?" Kaspersky Lab (January 5, 2017). https://www.kaspersky.com/blog/digital-depression/13781.

16. Clint Watts and Andrew Weisburd, "Can the Michelin Model Fix Fake News?," *The Daily Beast* (January 22, 2017). https://www.thedailybeast .com/can-the-michelin-model-fix-fake-news.

17. Elizabeth Dwoskin and Hamza Shaban, "Facebook Will Now Ask Users to Rank News Organizations They Trust," *The Washington Post* (January 19, 2017). https://www.washingtonpost.com/news/the-switch/wp /2018/01/19/facebook-will-now-ask-its-users-to-rank-news-organiza tions-they-trust/?utm_term=.b94f6b68f0fd.

18. Reid Standish, "Why Is Finland Able to Fend Off Putin's Information War?," *Foreign Policy* (March 1, 2017). http://foreignpolicy.com /2017/03/01/why-is-finland-able-to-fend-off-putins-information-war.

19. Jason Horowitz, "In Italian Schools, Reading, Writing and Recognizing Fake News," *The New York Times* (October 18, 2017). https://www.ny times.com/2017/10/18/world/europe/italy-fake-news.html.

20. Katherine Quinn, "Cornell Researchers Study Snapchat's Appeal," *Cornell Daily Sun* (March 15, 2016). http://cornellsun.com/2016/03/15 /cornell-researchers-study-snapchats-appeal.

21. Kayla Webley, "How the Nixon-Kennedy Debate Changed the World," *Time* (September 23, 2010). http://content.time.com/time/nation /article/0,8599,2021078,00.html.

EPILOGUE

1. Zack Beauchamp, "Trump's Republican Party, explained in one photo." Vox. August 6, 2018. Available at: https://www.vox.com/policy-and -politics/2018/8/6/17656996/trump-republican-party-russia-rather -democrat-ohio.

2. https://www.justice.gov/file/1035477/download.

3. https://www.npr.org/2018/02/17/586698361/the-russia-investigations -mueller-indicts-the-internet-research-agency.

4. https://www.justice.gov/file/1080281/download.

5. Ibid.

6. https://www.volkskrant.nl/wetenschap/dutch-agencies-provide -crucial-intel-about-russia-s-interference-in-us-elections-b4f8111b/.

7. http://fm.cnbc.com/applications/cnbc.com/resources/editorialfiles /2018/10/19/document-5.pdf.

8. https://www.cnbc.com/2018/10/19/woman-linked-to-russian-troll -farm-charged-with-interference-in-2018-midterms.html.

9. https://www.intelligence.senate.gov/press/new-reports-shed-light -internet-research-agency's-social-media-tactics.

10. https://www.washingtonpost.com/politics/russians-interacted-with -at-least-14-trump-associates-during-the-campaign-and-transition /2018/12/09/71773192-fb13–11e8–8c9a-860ce2a8148f_story.html.

11. https://www.nytimes.com/2018/11/29/us/politics/fact-check-cohen -trump-.html.

12. https://www.thedailybeast.com/russian-hackers-new-target-a -vulnerable-democratic-senator.

13. https://www.npr.org/2018/08/01/634696302/sen-jeanne-shaheen-says -she-was-a-target-of-a-hacking-attempt.

14. https://www.washingtonpost.com/technology/2018/08/04/several -groups-banned-by-facebook-had-strong-similarities-twitter-accounts -linked-russia-six-weeks-ago/?utm_term=.d73a60afd373.

15. https://www.thedailybeast.com/russian-troll-farm-internet-research -agency-has-new-meta-trolling-propaganda-campaign.

16. John Sipher, "Convergence Is Worse Than Collusion." *The Atlantic*. August 13, 2018. Available at: https://www.theatlantic.com/ideas /archive/2018/08/convergence-is-worse-than-collusion/567368/.

17. http://www.msnbc.com/rachel-maddow-show/why-did-trump-endorse
-old-soviet-talking-points-afghanistan.

18. https://www.cnbc.com/2018/08/08/rand-paul-delivers-trump
-letter-to-putin-administration-in-moscow.html.

19. https://newrepublic.com/article/110223/the-rise-russia-gun-nuts.

20. https://www.npr.org/templates/transcript/transcript.php?storyId=
647174528; https://www.nytimes.com/2018/08/04/us/politics/maria
-butina-nra-russia-influence.html; https://www.justice.gov/opa/press-re-
lease/file/1080766/download.

21. https://www.washingtonpost.com/local/public-safety/alleged-russian
-agent-maria-butina-had-ties-to-russian-intelligence-agency-prosecutors
-say/2018/07/18/a1a4042c-8a01-11e8-a345-a1bf7847b375_story
.html?utm_term=.e6ecc3ff9a8e.

22. https://www.bloomberg.com/news/articles/2018–12–14/meet-the
-russians-picked-for-butina-s-trip-to-trump-breakfast.

23. https://www.cnn.com/2018/02/07/politics/russia-delegation
-washington-prayer-breakfast/index.html.

24. https://web.archive.org/web/20140818073152/http:/archive.adl.org
/anti_semitism/duke_russia.html.

25. https://www.thedailybeast.com/richard-spencer-and-white-supremacists
-return-to-charlottesville-chanting-you-will-not-replace-us.

26. https://leagueofthesouth.com/российский-охват-russian-outreach/.

27. https://www.apnews.com/a4ffacbbf51748fc86eb442fb3db18b5.

28. https://www.washingtonexaminer.com/news/congress/gop-senators
-headed-to-moscow-to-smooth-things-over-with-putin.

29. https://dailycaller.com/2018/03/08/the-ever-changing-russia-narrative
-in-american-politics-is-cynically-false-public-manipulation/.

30. https://www.nbcnews.com/tech/tech-news/twitter-s-response-russia-
inquiry-inadequate-democratic-senator-says-n805646.

31. https://blog.twitter.com/official/en_us/topics/company/2018
/enabling-further-research-of-information-operations-on-twitter
.html.

32. https://blog.twitter.com/official/en_us/topics/company/2018/twitter
-health-metrics-proposal-submission.html.

33. https://medium.com/dfrlab/youtubes-kremlin-disinformation
-problem-d78472c1b72b.

34. https://www.facebook.com/facebookmedia/blog/working-to-stop
-misinformation-and-false-news.

35. https://www.nytimes.com/2018/11/14/technology/facebook-data
-russia-election-racism.html and https://www.nytimes.com/2018/11/29/
technology/george-soros-facebook-sheryl-sandberg.html.

36. https://www.warner.senate.gov/public/index.cfm/the-honest-ads-act.

37. https://www.politico.com/story/2018/03/22/election-security-bill
-congress-437472.

38. https://www.washingtonpost.com/technology/2018/11/08/white-
house-shares-doctored-video-support-punishment-journalist-jim-
acosta/?utm_term=.511d285663f2.

39. For further discussion of artificial intelligence, social media and its
trajectory, see https://www.washingtonpost.com/news/democracy-post/
wp/2018/09/05/artificial-intelligence-is-transforming-social-media
-can-american-democracy-survive/?utm_term=.3446fd4e2ff3.

40. https://www.wired.com/story/how-whatsapp-fuels-fake-news-and
-violence-in-india/.

41. https://www.politico.com/story/2018/03/22/election-security-bill
-congress-437472.

42. https://www.politico.eu/article/internet-governance-facebook
-google-splinternet-europe-net-neutrality-data-protection-privacy-united
-states-u-s/.

43. https://www.nytimes.com/2018/10/20/technology/politics-apps
-conservative-republican.html.

44. https://www.vanityfair.com/news/2018/08/steve-bannon-big-data
-facebook-twitter-google.

About the Author

CLINT WATTS is a Robert A. Fox Fellow in the Foreign Policy Research Institute's Program on the Middle East as well as a senior fellow at the Center for Cyber and Homeland Security at the George Washington University.